W9-AWU-887

First printing: August 2003
New updated and expanded version: January 2010
Second updated and expanded printing: May 2020

Master Books®, P.O. Box 726, Green Forest, AR 72638.

Master Books is a division of the New Leaf Publishing Group, Inc.

ISBN 13: 978-0-89051-581-5
ISBN 13: 978-1-61458-707-1 (digital)
Library of Congress Number: 2003106347

Cover Design: Diana Bogardus
Interior Design: Terry White

The New King James Version of the Bible is quoted unless otherwise indicated.

Printed in China

Photo Credits:

Shutterstock.com: Pages 2–3, 5, 8, 18–20, 22–23, 27, 29, 32, 35-36, 41–42, 44–46, 48–51, 53–55, 57–58, 60–63, 65–66, 68–69, 71–72, 74–75, 77–78, 80-84.

NASA: Pages 7, 9, 14–16, 21–23, 25–29, 33, 39, 42, 44–45, 47, 56, 59, 70, 72, 73, 75, 82, 86.

CONTENTS

EDUCATIONAL OBJECTIVES

Our Created Moon has been developed as an educational resource and can be utilized to assist in classroom study, independent learning, or homeschool settings. Utilizing the intellectual or cognitive section of Benjamin Bloom's Taxonomy, cognitive skills are developed from basic (e.g., one learning the alphabet) through advanced levels (e.g., one writing an opinionated research paper).

The following list displays the words used on each chapter objective and the level of skill desirable:

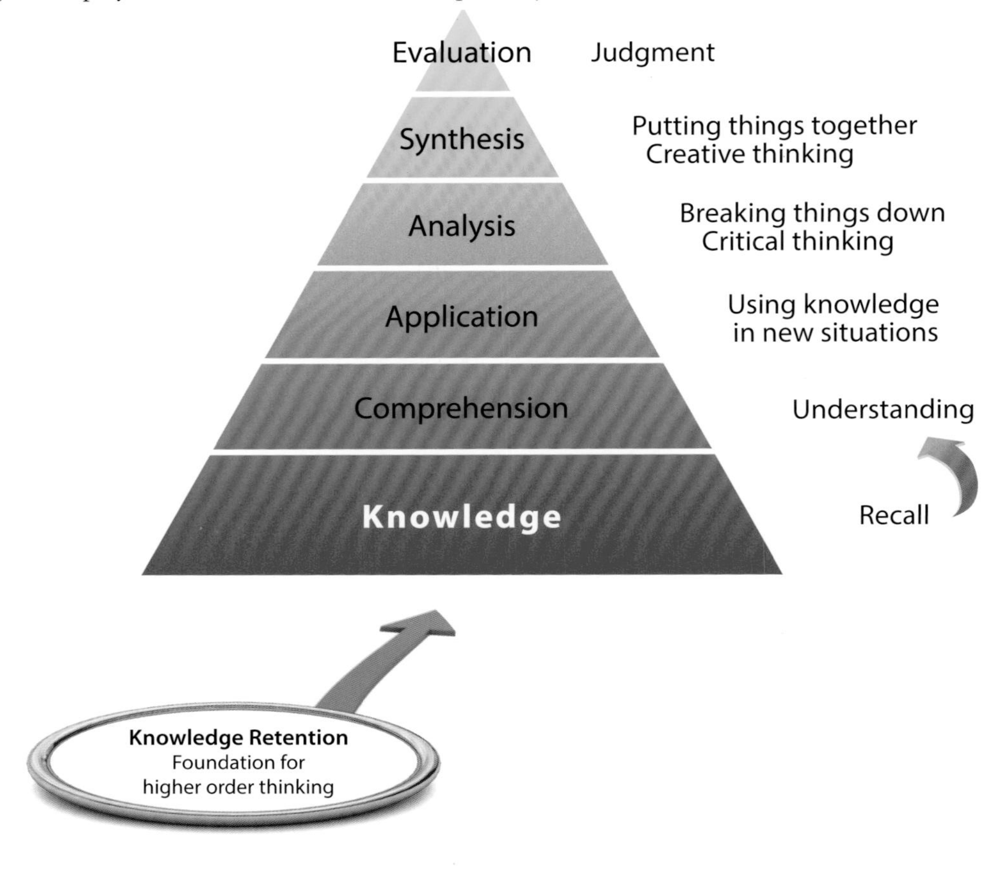

1. **Knowledge objectives**: focuses on a student remembering or recognizing basic facts, ideas, and concepts, using words such as define, describe, and identify.
2. **Comprehension objectives**: focuses on a student demonstrating an understanding of these facts, ideas, and concepts, using words such as cite, discuss, estimate, explain, and summarize.
3. **Application objectives**: focuses on a student taking basic knowledge or facts and applying it in new situations or different ways, using words such as articulate, assess, and relate.
4. **Analysis objectives**: focuses on a student examining information and finding a cause or motive behind any associated relationships, using words such as analyze, compare, contrast, differentiate, illustrate, and outline.
5. **Synthesis objectives**: focuses on a student combining different pieces of information, compiling them in ways that provide unique or alternative results, using words such as evaluate and respond.
6. **Evaluation objectives**: focuses on a student assessing differing ideas, making judgments concerning the value of these ideas and concepts, using words such as appraise and debate.

HOW TO USE THIS BOOK

Watch for the following highlights throughout the book to help provide the best educational experience:

Words to Recognize
At the beginning of each chapter, look over these words that hold special significance to the upcoming questions and answers. At the end of the chapter you can check the chapter word review to ensure that these terms were properly understood.

Learning Objectives
Examine these learning objectives at the beginning of each chapter. They were developed to help the teacher or independent learner build upon basic learning principles as a foundation for the more developed learning skills and are based on the question/answer format of each chapter.

Moon Activities
Make your own interesting findings and observations from suggested activities at the end of each chapter. Watch for this symbol denoting something that can be done to further enhance the learning experience and take the theory from the pages into your own lab or backyard.

Shooting for the Moon
To shoot for the moon is a phrase that suggests setting high goals for yourself. Watch for this symbol to find unique and fascinating facts and insights on the moon that are found throughout the text.

Flip Pages
The book has been designed so you can flip through the pages and watch the phases of the moon where the page numbers are found.

PREFACE

The first edition of this book was written in 1978. Since then, a wealth of new information has appeared concerning the moon and space. There have been lunar probes such as LCROSS, completion of the Hubble Space Telescope and the Keck Telescope in Hawaii, and even a new theory of lunar origin by a collision process.

In this book we concentrate on the topics that directly relate the moon to creation studies. This goal is both straightforward and rewarding because the moon is an excellent witness to creation in many ways. We use a question-answer format throughout the book, as well as added educational material for this new updated and expanded version. Special thanks to NASA engineer Tom Henderson for helpful suggestions.

For many years it has been the deep desire of the authors of this book to glorify the Creator of the world. Our greatest discovery, however, has been the grace and mercy of this awesome God, in providing eternal salvation through faith in His Son. The same Person who created the moon and the entire universe became a human being and died on a cross to pay the full penalty for sin that we could never pay. Such sacrificial love transforms all things for those who believe in Him. "If you confess with your mouth the Lord Jesus and believe in your heart that God has raised Him from the dead, you will be saved" (Rom. 10:9). "Praise Him, sun and moon. . . . Praise the LORD . . . kings of the earth and all peoples. . . . For His name alone is exalted; His glory is above the earth and heaven" (Ps. 148:3–13).

—Dr. Don DeYoung and Dr. John C. Whitcomb

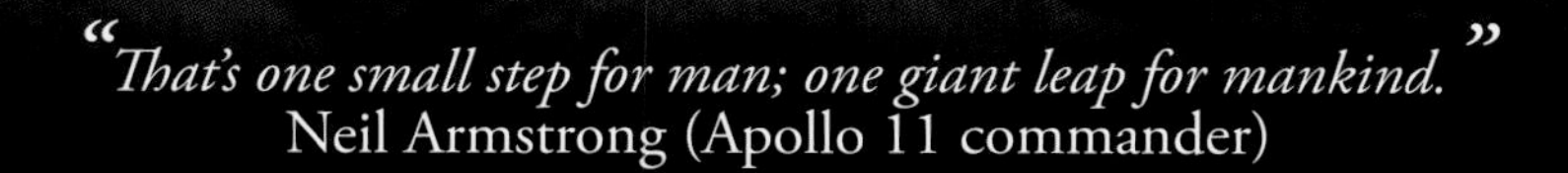

"That's one small step for man; one giant leap for mankind."
Neil Armstrong (Apollo 11 commander)

1 OUR NEAREST NEIGHBOR

Learning Objectives

1. Define what makes something a moon.
2. Describe the distance of the moon from earth, what keeps it in the sky, and its size.
3. Summarize why we see only one side of the moon, what caused the lunar craters, and the moon's surface features.
4. Explain what moon rocks are like, the notion of water on the moon, and the question of possible life on the moon.
5. Describe the cause of moon phases and how they affect the earth.
6. Analyze when eclipses occur and what causes ocean tides.
7. Discuss the Apollo program and what it was.
8. Compare our moon with other moons in the solar system.

Words to Recognize

breccias, cold traps, libration, lowlands, lunar eclipse, lunar highlands, moon, neap tides, regolith, sidereal period, solar eclipse, spring tides, synodic period

1. What is a moon?

A moon is any natural satellite that orbits a planet. It is held captive by the planet's gravity force and moves in a continuous elliptical orbit. Our moon circles the earth about once each month. There are about 100 known planetary moons in the solar system, and additional small moons continue to be found circling the outer planets.

Our moon is described as "the lesser light that rules the night" in Genesis 1:16. The actual word *moon* comes from Old English. In the Old Testament the most common Hebrew word translated "moon" is *hōdesh*, signifying the beginning of a new month or new moon. Another Old Testament word for moon is *ya-reah*, also related to the word *month*. Three times the Hebrew word used is *l^e^bana*, meaning "white one." In the New Testament, the Greek word for the moon is *selene*, meaning "brilliant" or "attractive." Our beautiful, dependable moon lives up to its name.

Table 1–1. The relative nearness of the moon can be appreciated by comparing its distance from Earth with several other space objects:

	Moon	238,712 miles (384,090 km)
	Sun	93,000,000 miles (149,637,000 km)
	Pluto	3,700,000,000 miles (5.95 billion km)
	Alpha Centauri	24,000,000,000,000 miles (38.6 trillion km) The nearest nighttime star, 4.3 light years away
	Big Dipper stars	558,000,000,000,000 miles (898 trillion km) An average of 100 light years away
	Remote Galaxies	55,000,000,000,000,000,000,000 miles (88.5×10^{21} km) 10 billion light years away

Other World Travel

Our moon is the only other "world" that astronauts have visited. From the desolate lunar surface the earth appears as a pleasant blue oasis in space.

2. How far away is the moon?

Our moon is the nearest neighbor to Earth in space. Its distance varies annually between 225,600 and 251,815 miles, with an average of 238,712 miles (384,090 km). There are many alternate ways to express this Earth-moon separation:

- 238,712 miles (center to center)
- 384,090 km
- a light travel time of 1.3 seconds
- a three-day trip for Apollo astronauts
- at 60 mph the trip would take 52 months
- equal to ten trips around the world
- 400 times closer than the sun

3. What keeps the moon in the sky?

How can a massive round rock weighing 81,000,000,000,000,000,000 tons (7.35×10^{22} kg) float overhead in our sky? Actually, it doesn't; the moon falls toward the earth continually. However, it does not get any closer to us. This paradox occurs because of the moon's orbital motion. Figure 1-1 shows how the moon's tangent speed and falling motion add together to result in a smooth, curving orbit around the earth. The explanation is similar for all orbiting objects, including moons, artificial satellites, and planets. If the moon's tangent speed ceased, it then would fall directly toward the earth and collide with us. On the other hand, if gravity ceased, the moon would leave its Earth orbit on a straight line path like a stone from a whirling slingshot.

The moon moves about 3,350 feet (1,020 m) each second tangentially while falling just one-sixteenth inch (1.6 mm) toward the earth. This combination results in a smooth, elliptical orbit. The attractive force toward the earth, called the centripetal ("center-seeking") force, is provided by gravity. Gravity is an attractive force occurring between all objects, whether on earth or in space. For example, your weight is simply the gravity attraction between you and the earth. On the scale of large space objects the gravity force becomes immense. The attraction between the earth and moon is about 4.5×10^{19} pounds, or over 20 million billion tons (2×10^{19} kg). This gravity force continually acts on the moon to maintain its orbit like a whirling ball held by a string, and, in turn, pulling back on the earth, the moon's gravity force results in our ocean tides.

What really is gravity? This mysterious force continues to puzzle scientists even as it gives stability to the entire universe. How is gravity able to act across vast stretches of empty space, and why does it exist in the first place? Science has never been very successful in explaining fully such natural phenomena as gravitation. Einstein proposed in 1915 that the gravity force results from a "distortion of the fabric of space." Others have searched for unseen graviton particles that move between space objects. Whatever the case, clearly this universal force rule cannot somehow arise by gradual change such as biological mutation or natural selection. Gravity was established from the very beginning of time. Gravity, along with every other intricate physical law and constant, is surely a testimony to God's planned, orderly creation.

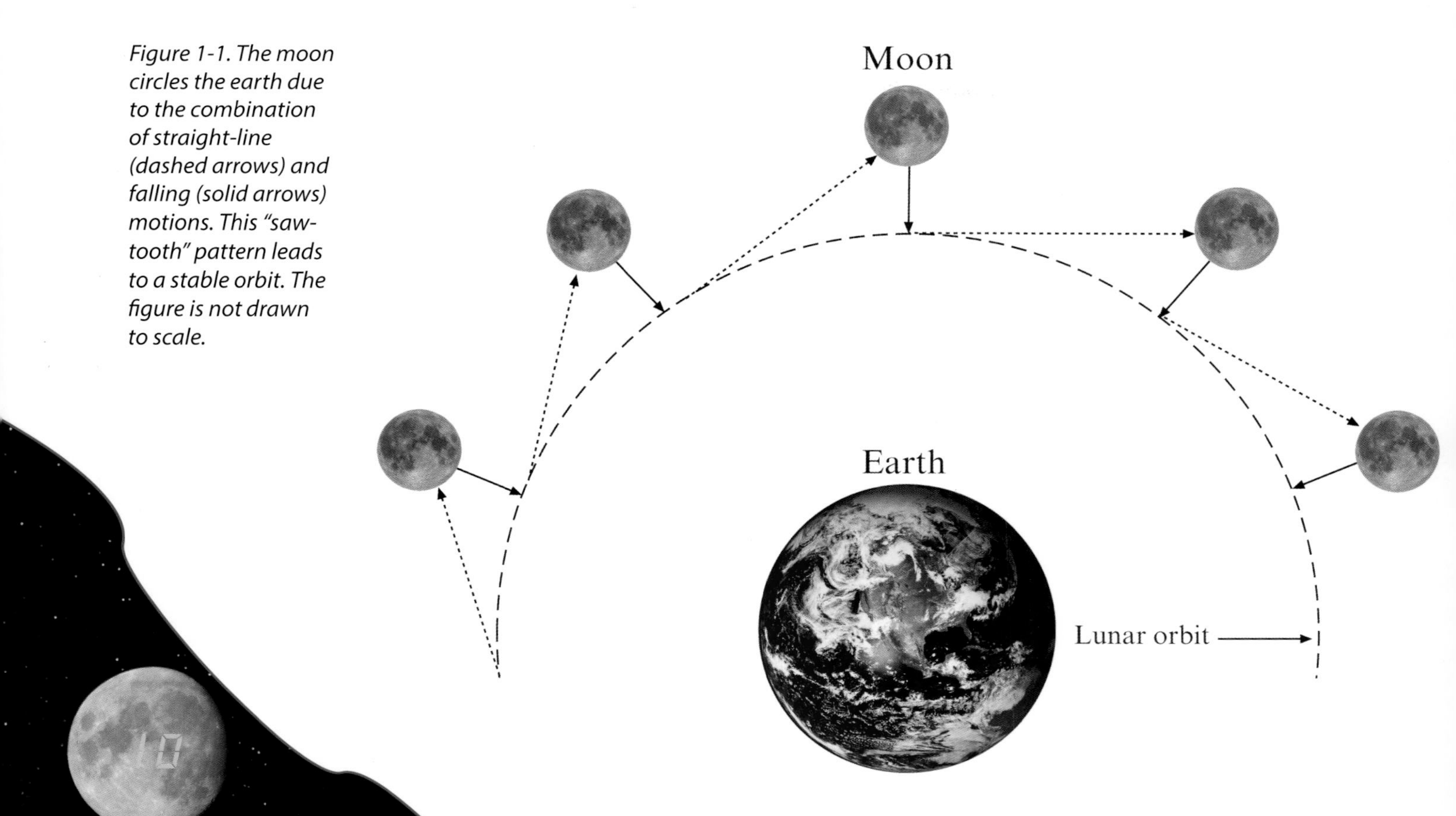

Figure 1-1. The moon circles the earth due to the combination of straight-line (dashed arrows) and falling (solid arrows) motions. This "saw-tooth" pattern leads to a stable orbit. The figure is not drawn to scale.

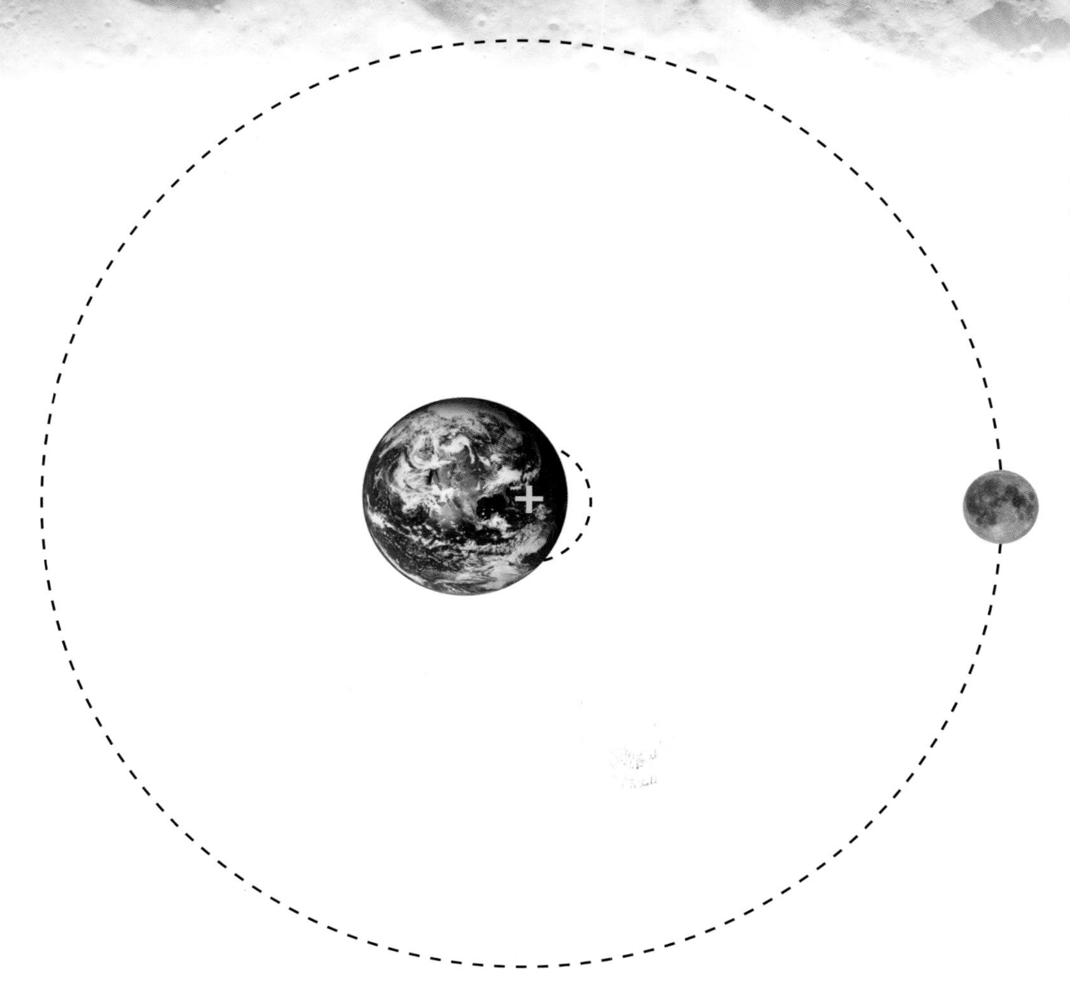

Figure 1-2. The motions of the earth and moon about their mutual center of mass (+). The earth's center moves around the small dotted circle as the moon moves around the much larger dotted path. The figure is not drawn to scale.

4. How large is the moon?

Our moon is approximately one quarter the size of the earth. The respective diameters are 2,160 miles (3,475 km) and 7,928 miles (12,756 km). In terms of actual mass, the earth is 81 times more massive than the moon. It is for this reason that we say the moon orbits the earth. Actually, both objects move about their mutual balance point or center-of-mass. This balance point, or *fulcrum*, is positioned within the heavy earth, which moves on a much smaller circle than the moon (Figure 1-2). The actual motion is much more complicated than this because of gravity interactions with the sun and other planets. The center-of-mass position changes continually depending on the positions of all these objects. A moment-by-moment analysis is called *a many body problem* in physics, and remains a difficult challenge even for the most powerful computers available.

Our moon is very large when considering the size of the earth. There are four other solar system moons larger than ours, but they are typically much smaller than their planets (Figure 1-3). Only Pluto's moon, Charon, is closer in size to its planet than the earth-moon system. All the other moons are less than 5 percent of the diameter of their planets. This makes the lunar masses less than 0.025 percent that of their planets. For this reason, our moon is sometimes called a "secondary" or "double planet" companion to the earth. Suppose the moon remained at its current distance but its mass was reduced 100 times to make it an "average size" solar system satellite. As a result, the moon's diameter would be decreased by about 78.5 percent. Since the light reflected from the moon depends on its surface area, which is proportional to the square of its diameter, the full moon brightness then would be reduced to just 5 percent of its present value, a 20-fold diminishment of its light. The moon's unusually large size, therefore, is necessary for it to provide significant evening light.

The Book of Genesis states, "Then God made two great lights: the greater light to rule the day, and the lesser light to rule the night" (Gen. 1:16). The Hebrew term translated *light* in this passage is flexible enough to include light reflectors such as the moon and the planets. Some critics have called Genesis 1:16 untrue because the sun is not the largest star in all of space, and the moon also produces no light of its own. However, the fact that the moon, which is a reflector, and the sun, which is far from being the largest star, are named "two great lights" is perfectly consistent with the language of appearance that the Bible uses throughout. A God who could not communicate with men in terms that they could understand would be limited indeed. From our human perspective, the only light that truly dominates the night sky is the nearby moon.

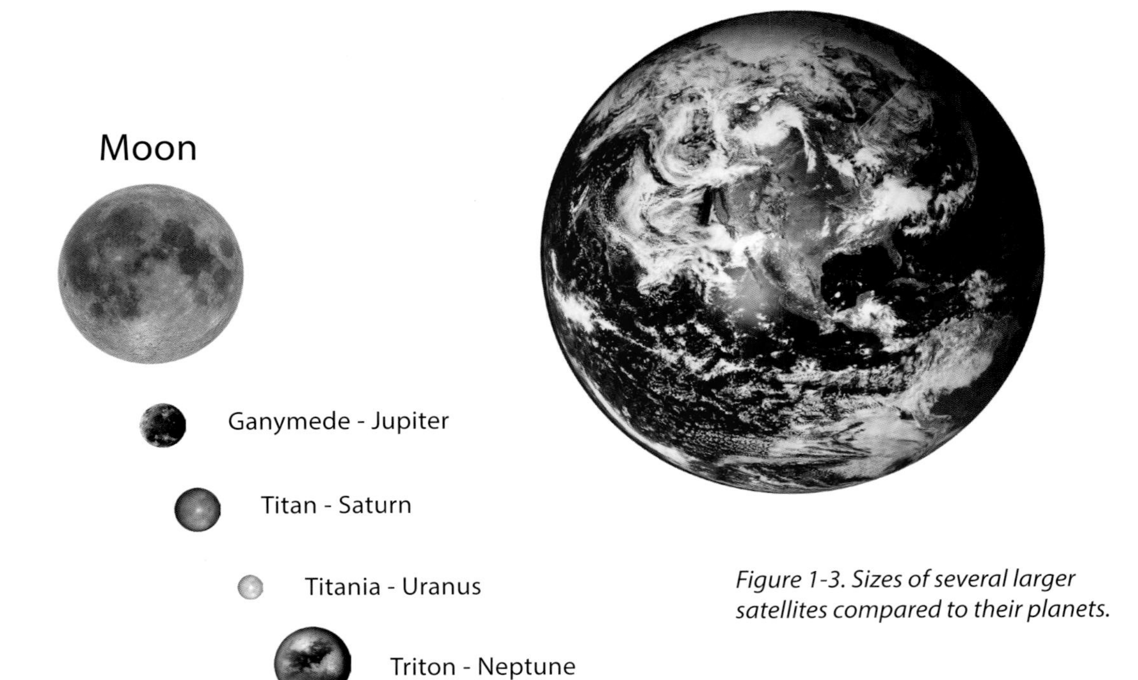

Figure 1-3. Sizes of several larger satellites compared to their planets.

5. Why do we see only one side of the moon?

The moon rotates once on its axis during the very same time that it orbits the earth, 29½ days (actually 29 d., 12 hr., 44 min., 2.98 sec.). As a result we always see the same side of the moon. This is not unusual for solar system moons; many others are likewise "locked in" as they orbit their respective planets. Our own moon has slightly more mass distributed on its near side so gravity attracts and holds this side toward the earth. The effect is similar to whirling a ball on a string with the side with the attached string always facing inward. The moon also "rocks" slightly back and forth during its orbit, which allows us to see slightly more than half the moon, about 59 percent. This is called *libration*.

Mark Twain in his novel *Pudd'nhead Wilson* (1894) made an interesting reference to the moon. He wrote, "Everyone is a moon and has a dark side which he never shows to anybody." This quote may well describe Twain's own struggle with his sinful nature. However, Mark Twain was incorrect in his description of the opposite side of the moon. It is dark only half the time and is fully lit during the new moon phase. The back of the moon is the *hidden* side but not always the *dark* side.

The opposite side of the moon may become an important future location for astronomers. Radio wave signals from space are a valuable source of information. They are detected with large radio telescope antennas. However, radio *noise* from satellites and Earth electronics can cause serious interference. A radio observatory constructed on the far side of the moon would be shielded from this interference. The lack of a lunar atmosphere could also be a bonus for clear observing of space.

Further comment is needed on the moon's rotation time. The stated value of 29½ days is measured with respect to the earth, called the *synodic* period (Figure 1-4). This is the time from one full moon to the next. The moon's rotation time is somewhat shorter with respect to the stars, 27½ days. This is called the *sidereal* period. Both lunar rotation times are commonly found in the literature.

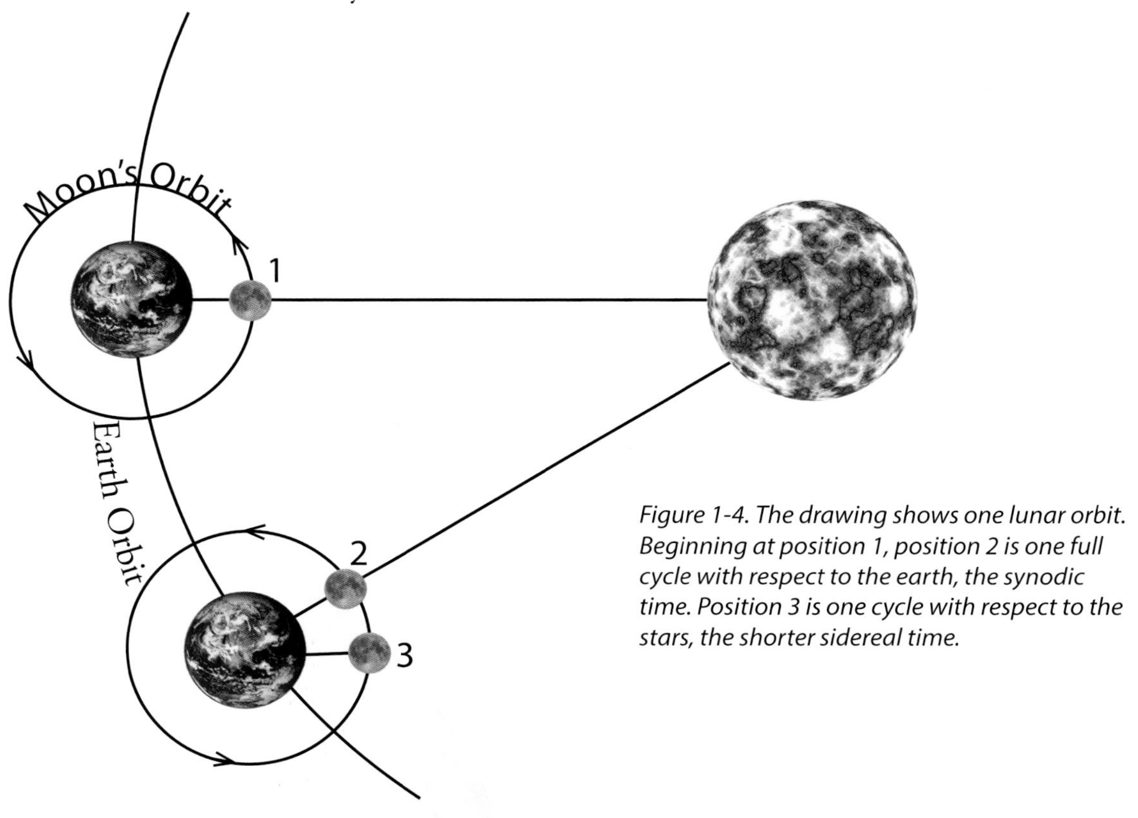

Figure 1-4. The drawing shows one lunar orbit. Beginning at position 1, position 2 is one full cycle with respect to the earth, the synodic time. Position 3 is one cycle with respect to the stars, the shorter sidereal time.

6. What caused the lunar craters?

Until recent decades this question often led to lively debate. Suggestions included volcanic activity, giant gas bubbles rising to the moon's surface from the interior, and collapsed sinkhole formations from dissolved underlying bedrock. The idea of impact collisions first became popular in the 1950s. Around this time, it was realized that a large meteorite collision had formed the well-known Barringer Crater ("Meteor Crater") in northern Arizona.

The moon displays countless numbers of craters. Some are hundreds of miles in size while most are much smaller. The Sea of Storms (*Oceanus Procellarum*), thought by some to be a crater remnant, exceeds the entire Mediterranean Sea in area. There are estimated to be 200,000 craters across the moon with diameters larger than a kilometer. Smaller craters within larger craters cover the moon like a battlefield. Since the moon has no atmosphere, all approaching space rocks strike the moon's surface. On Earth, in contrast, smaller objects may "skip off" our atmosphere and miss us completely, or else may burn up from friction on their rapid descent. Many other meteorites that hit the earth leave no trace because they land in the seas, which cover over two-thirds of the earth's surface.

About 200 craters have been identified across the earth. These blemishes gradually are eroded away by wind, water, and tectonic processes. In contrast, on the moon, craters become permanent records of past collisions.

7. What are the moon's surface features?

Suppose you could visit the moon to explore its surface. A space suit would be essential for your survival since there is no air to breathe in the near-perfect vacuum. With no air pressure, gas bubbles would form quickly in the bloodstream of an unprotected person. This dangerous situation is called decompression sickness or "the bends." If you land on the sunlit side of the moon, with its temperature up to 266°F (130°C), a reflecting suit with built-in air conditioning is essential. The lack of significant water or atmosphere results in a daylight temperature greater than the boiling point of water. Meanwhile, on the darkened portion of the moon, the temperature plunges lower than -292°F (-180°C).

You will notice shadows on the moon, and greatly reduced gravity. In fact, your weight will be only one-sixth of its value on Earth. For example, if you weigh 150 pounds on Earth, your moon weight will be just 25 pounds. The feeling is somewhat like the buoyancy of water. One can imagine the daily weather report for the moon: *no rain is in sight; also no breeze, clouds, sounds, or blue sky. Skies will be black instead. Humidity will remain at zero percent. The high temperature today will be a scorching 266°F. When night finally comes after two weeks of sunlight, the temperature will plunge to -292°F.* Lunar astronaut Buzz Aldrin called the moon a place of "magnificent desolation." How different the moon is from the specially prepared earth.

During the imaginary moon visit you will notice large flat areas or *lowlands* that cover one-half of the moon's visible side. Early astronomers named them *maria* (mar-eé-a, singular: mare), the Latin word for seas. They are indeed seas, but made of hard basalt rock instead of water. From the earth, these hardened lava flows have the appearance of large, circular, dark-colored patches. *The lunar highlands* are rugged mountain ranges that appear from Earth as light-colored patches. Some of these lunar peaks rise five miles above the surrounding plains, rivaling Earth's Mount Everest in height. Especially high elevations near the northern lunar pole, the so-called mountains of eternal light, withdraw from constant sunshine only at times of a lunar eclipse. Astronomers do not understand how the lunar mountains formed. The moon does not show mountain-forming activity analogous to earthquakes, seafloor spreading, or continental drift. Creationists might suggest that the moon was created with great variety in its topography, similar to the earth. Appendix 1 suggests some particular lunar features of special interest for close observation.

Besides the lunar mountains, basaltic seas, and craters, many other distinctive surface features of the moon can be seen with a telescope. Bright *rays* appear to radiate outward hundreds of miles from some of the larger craters. They are debris tracks deposited from the *ejecta* of major impact collisions. The rays appear as elongated, gentle hills rounded by further impact erosion. Sinuous lunar *rilles* that meander across the moon's surface are probably collapsed lava drainage channels, since the lack of water rules out river channels. The many visible cracks and valleys on the surface are apparently adjustments in the moon's surface to stress from heating and tidal pull. The far side of the moon was first photographed in 1959 by the Russian spacecraft *Luna 3*. Fanciful speculation about hidden features and even a lunar civilization was ended by pictures of emptiness. There are many large craters and seas of solid basalt.

Rocks and large boulders are strewn about the lunar surface. But there are no rivers, lakes, grass, or trees to complete the landscape. Upon returning from a voyage to the moon, one conclusion is inescapable: there is no place like home!

8. What are moon rocks like?

The Apollo astronauts returned to Earth with about 843 pounds (382 kg) of moon rocks, core samples, pebbles, sand, and dust for study. Some meteorites collected on the earth appear to have been blasted from the moon by past impacts. The lunar rocks collected by astronauts resemble Earth varieties in some respects and differ significantly in other ways. The three varieties of collected samples are *crystalline rock, soil,* and *breccia*. The geological term *regolith* is given to the general lunar surface collection of dust, pebbles, and boulders.

The lunar *crystalline rocks* contain the same chemical minerals found in Earth rocks. Lunar basalt is common, similar to our abundant terrestrial volcanic rock. It forms by cooling from molten lava, and is especially common in the lunar maria areas. The small crystals within the basalt suggest a rapid cooling of the moon's surface in the past. Some of this material also may be original, created lunar crust. All the moon rocks contain higher proportions of heat-resistant elements such as calcium, aluminum, and titanium than Earth rocks. Conversely, the easily vaporized elements sodium, potassium, and lead are relatively depleted on the moon.

Moon rock presented to Smithsonian Institute by the Apollo 11 crew, 1969.

Along with basalt, another important lunar rock is *anorthosite*. It is light gray in color, which makes the lunar highlands lighter in appearance than the basalt-covered maria. The color contrast between anorthosite and basalt is also responsible for the appearance of the illusion of a human face ("man-in-the-moon"), which some people see on the moon's near side.

The lunar *soil* consists of the powdered remains of many collisions between meteorites and the igneous surface. Unlike the typical soil of Earth, it contains no organic matter and very little moisture. Small, bright beads of colored glass give variety to the soil, indicating the melting of material during past impacts and subsequent rapid cooling.

Breccias are rocks composed of small rock fragments, glass, and soil that have been compacted into cohesive rocks. The lunar particles are sharply angular rather than rounded, as is the case for most terrestrial conglomerate rocks. Lunar breccias may result from *shock melting* during the impact of meteorites.

The moon lacks some of the most common rocks found on Earth, including granite, and also sedimentary varieties such as limestone, shale, and sandstone. The lack of sedimentary deposits rules out any hint of a large lunar water supply.

9. What is the significance of possible water on the moon?

Most Earth rocks contain at least 1 to 2 percent water by total weight, locked within their crystal structure. Some researchers have suggested that vast deposits of water in the form of ice might exist in isolated, protected spots near the moon's poles, perhaps within deep craters. Such locations are in permanent shadow and are called *cold traps*. The water supposedly was delivered to the moon by comets or meteorites in the distant past. However, a search by the lunar probes Clementine and Lunar Prospector (1999) failed to verify lunar ice.

There is a common assumption in astronomy that the presence of water anywhere in space will likely lead to evolved life. This has led to an intense search for water on the moon, Mars, and elsewhere, all without success thus far. Water reserves may eventually be found in space, perhaps on some of the other solar system moons. But the presence of life is another question entirely. The assumption is completely false that one needs to "just add water" to begin life forms. Living things, whether plant or animal, large or small, are not quite that simple!

One cannot help but reflect on the extreme differences between the comfortable earth and its important but entirely inhospitable satellite companion. If the earth were without its protective atmosphere and abundant water supply, the terrestrial environment would be very similar to that of the moon. The extreme properties of the moon, and all of the other planets, provide a vivid demonstration of God's care in providing a beautiful earth.

10. Is there life on the moon?

The short answer to this question is no. During the Apollo voyages to the moon there were fears of possible infecting microbes living in the unknown lunar environment. Some scientists suggested that these foreign invaders might decimate life across the earth. As a precaution, the first astronauts and their treasure of returned rock samples were temporarily placed in isolation chambers upon returning to the earth. However, the lunar rocks soon were determined to be barren of life, and the quarantine period was dropped completely following the third Apollo landing. The lack of significant traces of lunar water and life was a disappointment to those who had hoped the moon would support the Darwinian concepts of the spontaneous origin and evolution of life. The lack of life-building carbon and free oxygen also quenched evolutionary optimism concerning the moon. Returned lunar rocks contain only "parts per million" traces of carbon and carbon compounds, which may have been due to contamination by the astronauts themselves.

APOLLO 11 LUNAR MODULE

If life *is* eventually found on the moon, on Mars, or elsewhere in space, there are only three possible sources: spontaneous evolution or abiogenesis, supernatural creation, or contamination from the earth. The first source, an evolutionary origin, is ruled out by its zero probability of occurrence (Moorehead and Kaplan, 1967), and by the failure of every origin of life experiment from the time of Louis Pasteur (1822–1895) until today. The second possible source, the creation of life in space, has no biblical support. Instead, the Bible clearly emphasizes the priority and uniqueness of life on Earth (Isa. 45:18). The third possible source of life in space is contamination. This is a reality in the vicinity of Earth, at least on the microscopic scale. We have soft-landed spacecraft on the moon, Venus, Mars, Jupiter, and elsewhere, and Earth microbes, including bacteria, inevitably rode along as passengers. Some of them might be able to survive in their present hostile environments.

Phase	Rises	Overhead	Sets
New	6 a.m.	Noon	6 p.m.
First Quarter	Noon	6 p.m.	Midnight
Full	6 p.m.	Midnight	6 a.m.
Third Quarter	Midnight	6 a.m.	Noon

Table 1-2. The approximate rising, overhead position, and setting times for the major moon phases. "Overhead" refers to the time when the moon is highest in the sky, not necessarily straight up.

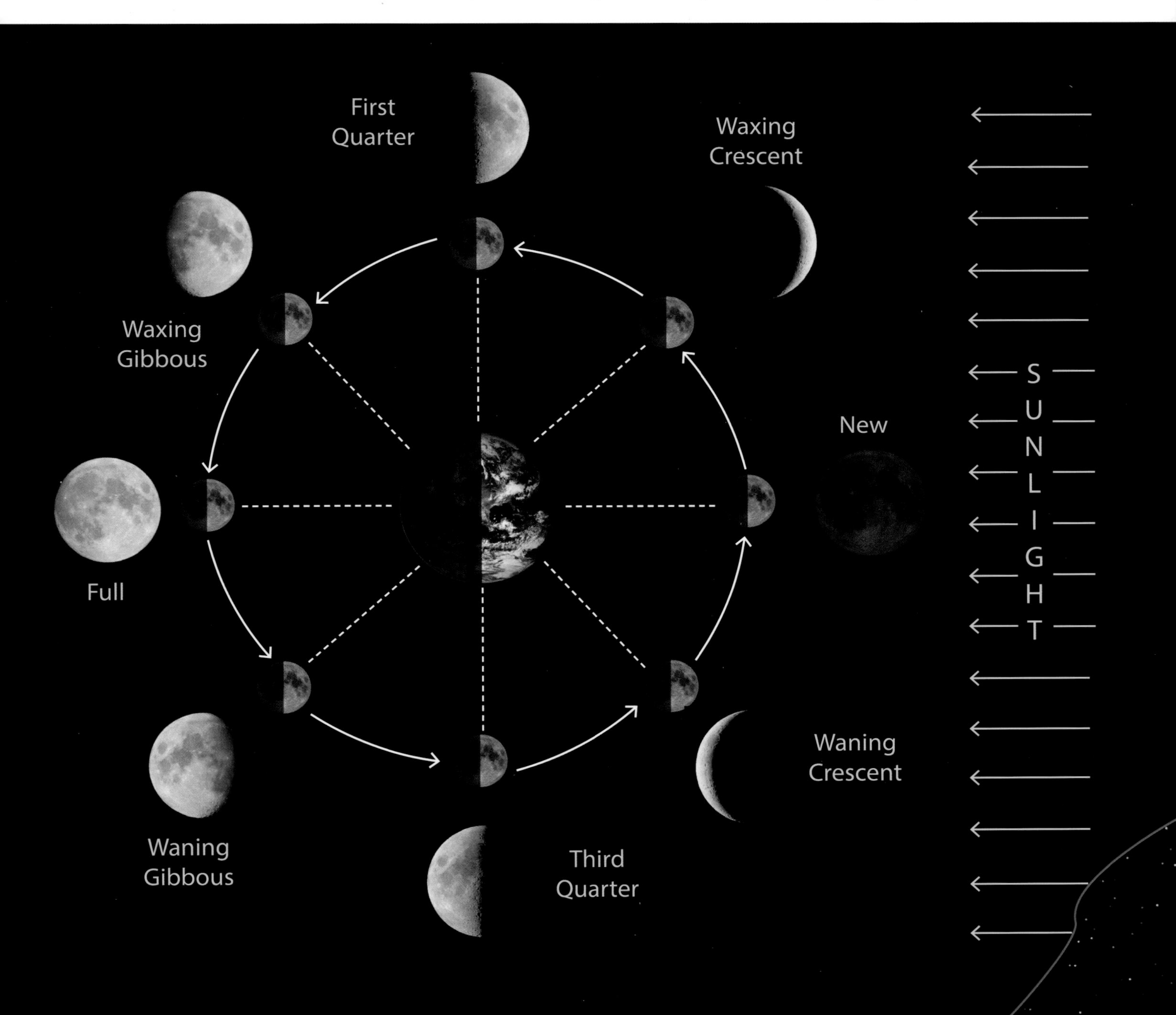

Figure 1-5. The phases of the moon with the corresponding positions of the earth, moon, and sun. The outer figures show the appearance of the moon as seen from the earth.

11. What causes moon phases?

The changing phases of the moon are of unending beauty and interest. They result from our partial view of the sunlit portion of the moon. The phases are *not* due to the earth's shadow, which only covers the moon during a lunar eclipse. Figure 1-5 shows the moon's changing appearance as it circles the earth. With a revolving period of 29½ days, the major phases — new moon, first quarter, full moon, third quarter — each occur about one week apart. The crescent and gibbous phases are also shown in the figure. Table 1-2 gives the times for the rising, highest position, and setting of the major moon phases.

The regularity of moon phases provides the basis for the lunar calendar discussed in chapter 3. The sighting of the waxing crescent phase was of special interest to Old Testament Jews in beginning each new month. This was called the new moon but was actually the thin sliver that occurs one to two nights following the invisible new moon phase. The new moon is mentioned in Psalm 81:3, Isaiah 1:13, Ezekiel 46:1, Hosea 2:11, and elsewhere. The return of the moonlight after its temporary disappearance was celebrated with festivals, sacrifices, and prayers. The waxing crescent phase is first observed in the western sky just after sunset, and watchmen were posted to look for this phase. When first seen each month, the news was sent throughout Israel from the Mount of Olives by beacon fires on distant hilltops. Other Old Testament festivals such as Passover and Sukkoth were celebrated at the time of the full moon.

12. Do moon phases affect the earth?

The full moon phase has been credited with influencing society in many ways. These include the birth rate, crime rate, mental health problems, and even stock market trends. However, statistical studies show little or no correlation of social behavior with the full moon phase.

An actual influence of moon phases concerns the tides. Tides are highest during the full moon and new moon phases. At these times, about two weeks apart, there is a lineup between the sun, earth, and moon, and the gravity of both objects then contributes to the tidal pull on the earth, called *spring* tides. These unusually high tides are not related to the spring season. The ocean tides are about two-thirds due to the moon because of its closeness, and one-third due to the sun's gravity. During the first and third quarter moon phases, the sun and moon are 90 degrees apart, and Earth tides then are lower, sometimes called *neap* tides (Figure 1-5).

Moon phases also have a slight influence on the weather. Satellite data shows a minor warming of the earth's polar regions by about one degree during the full moon phase (Pearce, 1997). This temperature change results from the increase in evening moonlight. Farmers also describe the influence of moon phases on planting, harvesting, and the handling of animals. Such effects, if real, are not understood scientifically, and they deserve further study.

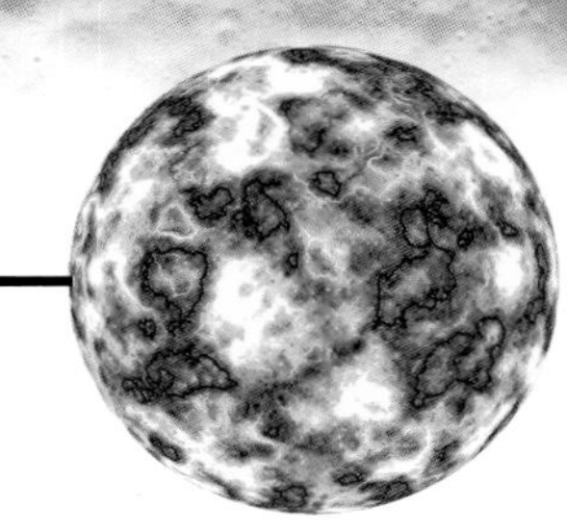

Figure 1-6. The tilted orbit of the moon compared with the earth's solar orbit.

13. When do eclipses occur?

A *lunar eclipse* occurs when the earth is lined up exactly between the sun and moon. This happens once or twice a year during the full moon phase (Figure 1-5). The eclipsed moon turns a red-brown color as it moves directly through the earth's shadow. The lunar eclipse lasts for about an hour and is not a particularly rare event. When the *new moon* moves exactly between the earth and sun, a *solar eclipse* occurs. Partial solar eclipses are common, but a total eclipse is very rare for a given location on the earth. A total solar eclipse occurs *somewhere* on Earth about three times each decade. A particular city might expect to witness a total solar eclipse only once about every 360 years. The event brings darkness to a small area and lasts for only a few minutes. During this time the temperature drops, birds begin to roost, and stars may appear. Lunar and solar eclipses do not occur every month because the moon is usually positioned above or below the plane of the earth-sun system. The moon's orbit around the sun is tilted 5° to the earth's solar orbit. Furthermore, the moon orbit slowly *precesses* or wobbles due to the sun's gravity pull. As a result, the sun, earth, and moon seldom line up exactly to cause lunar and solar eclipses (Figure 1-6).

How can the moon completely cover the sun during a solar eclipse? It would seem to be impossible since the moon is about 400 times smaller in diameter. However, the moon is also about 400 times closer to us than the sun. As a result, the sun and moon have the same apparent size in the sky, and are able to exactly eclipse each other. Computer studies show that this phenomenon is unique among the known moons of the solar system. Other moons provide at best a partial eclipse or else a lengthy total eclipse of the sun for their respective planets. This fortuitous size-distance balance between our moon and the sun is usually described as a "surprising coincidence" or a "lucky accident" in nature. However, this phenomenon actually points to yet another detail of design in the moon's creation for man's benefit and God's glory (chapter 3).

There will be special opportunities in coming years for many Americans to view a total eclipse of the sun. Three dates of eclipse totality across North America are:

August 21, 2017
April 8, 2024
August 12, 2045

Figure 1-7 shows the paths of North American total solar eclipses during 2001–2050. Temporary darkness occurs along the paths shown, with a width of several miles. Further from the center of the path, a partial solar eclipse is seen. The dates are lifetime opportunities for millions of people to witness the temporary loss of sunlight in the moon's dark shadow.

The first eclipse date listed, August 21, is a Monday. The eclipse touches the Northwest coast at about 10:15 a.m. (Pacific Time). The shadow then races across the United States at over 1,050 miles per hour (1700 km/hr). It passes over such major cities as St. Louis, Kansas City, Nashville, and Charleston.

The eclipse leaves the Southeast at 2:50 p.m. (Eastern Time), crossing the entire country in about 1½ hours.

Cloudy weather conditions sometimes diminish the dramatic view of a total solar eclipse. Seasoned eclipse chasers prepare for last-minute road travel to a clear site, in case clouds appear. Professional astronomers often charter aircraft to fly above clouds and collect data.

During the 2017 event, cameras and telescopes were set up all across the country along the eclipse path. However, one does not need high- tech equipment to enjoy a total solar eclipse. There are precautions needed, of course, when looking in the direction of the sun. A total eclipse lasts for just minutes, with blinding sunlight before and after. Still, the unusual midday darkening is a very moving experience for all observers. The majesty and clock-like precision of the created heavens catches everyone's attention, if only for a day.

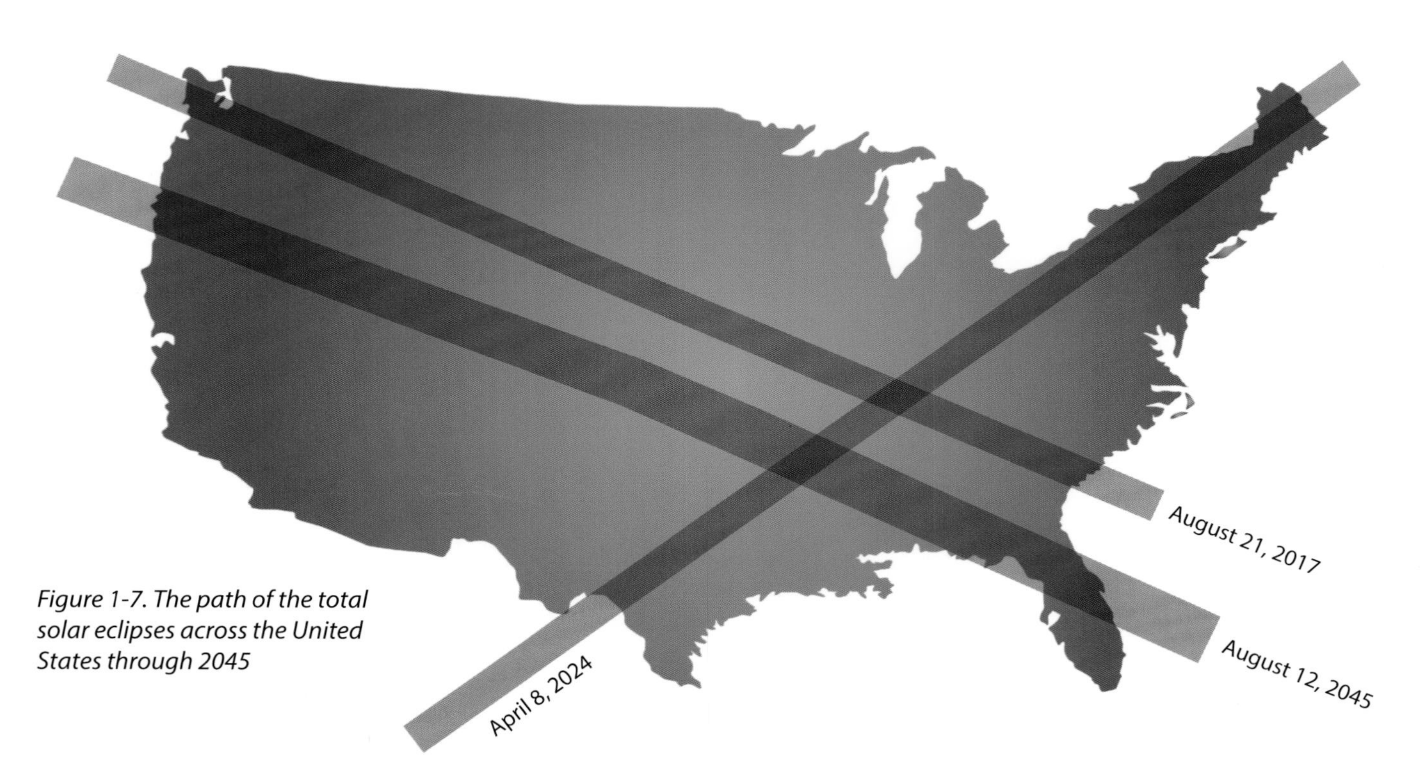

Figure 1-7. The path of the total solar eclipses across the United States through 2045

14. What causes the ocean tides?

Earth's tides result mainly from the moon's gravity pull with a smaller component, about one-third, caused by the sun's gravity. Tides actually depend on the change of gravitational force between opposite sides of the earth. Two tidal bulges continually occur on the earth due to the pull of the moon (Figure 1-8). The side of the earth nearest the moon feels the greatest gravitational pull and bows slightly outward in a high tide. The far side of earth, feeling the least gravity attraction, bows slightly *away* from the moon to produce another high tide. As the moon orbits the spinning earth, two high tides (and also two low tides) occur during approximately each 24-hour period at a given location on the earth. Actual tidal phenomena in any local region depend on the shoreline and ocean basin configurations. Due to the moon's orbit, the time of high tide at a given location shifts progressively about 50 minutes later each day. At the earth's equator, the tidal bulges move eastward at a speed of about 1,000 miles per hour.

Besides seawater, the land also feels the moon's tidal pull. When the moon is overhead, the crust of the earth rises by several inches. This occurs because there is a slight flexibility for expansion and contraction of the earth's rocky base. Tides also occur in the earth's upper atmosphere. This blanket of air expands and contracts in the sky with the moon's changing pull.

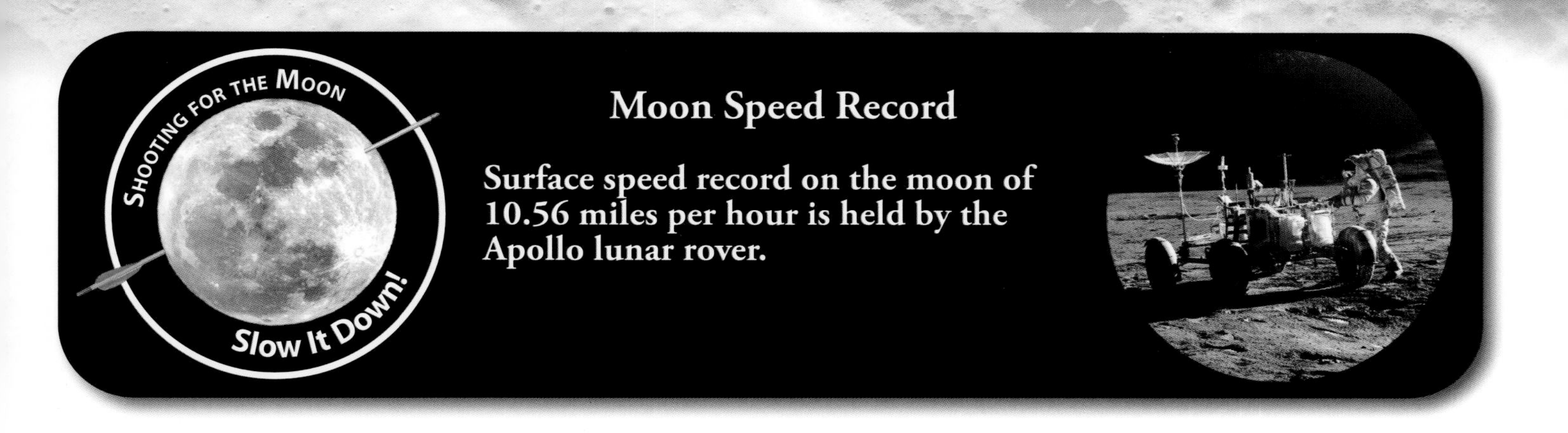

Moon Speed Record

Surface speed record on the moon of 10.56 miles per hour is held by the Apollo lunar rover.

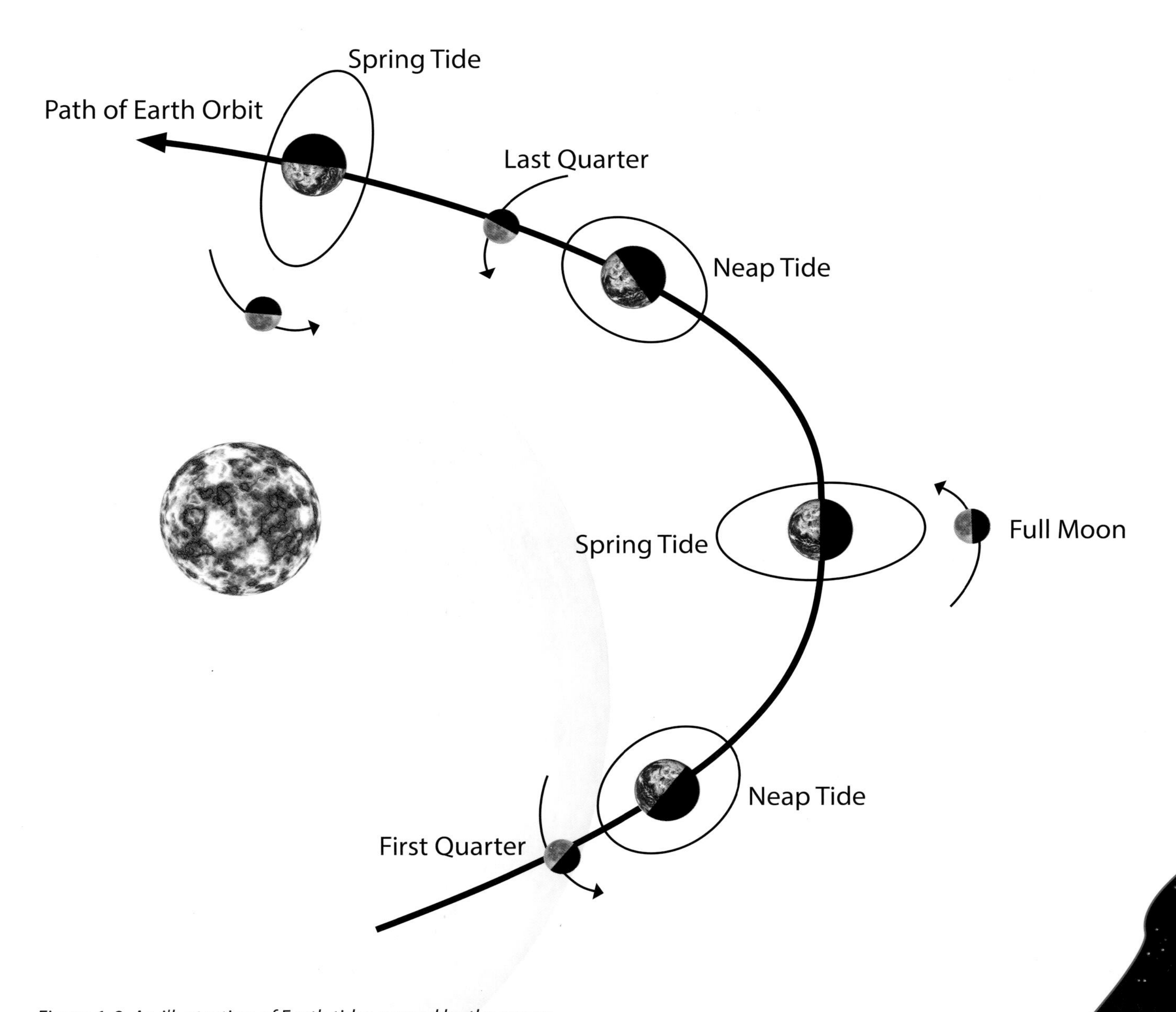

Figure 1-8. An illustration of Earth tides caused by the moon, appearing on both the near and far sides of earth, not drawn to scale. Spring tides about two weeks apart bring a maximum height of water. Neap tides have a minimum effect on the oceans.

15. What was the Apollo program?

During the 1960s and 1970s, major efforts were directed toward moon exploration. There was intense technological competition between the United States and Russia during this *cold war* era. After remaining beyond our grasp through history, the moon, in 12 short years, experienced landings by 21 Soviet and 13 American unmanned spacecraft. In a famous 1961 speech, President John F. Kennedy set a national goal of manned moon exploration before the end of the decade. Tragically, he was assassinated in 1963, but his lofty goal was reached. The resulting Apollo program delivered six manned crews aboard Saturn rockets to the moon's surface. The mighty three-stage Saturn rockets were about 360 feet long with engines generating 7.5 million pounds of thrust at liftoff.

Astronaut Neil Armstrong took his first dramatic footsteps upon the Sea of Tranquility on July 20, 1969. Altogether, a dozen astronauts visited the lunar surface, covering 60 miles and conducting dozens of experiments. Figure 1-9 shows the Apollo landing sites on the moon. At its peak, the American space program supported 400,000 technical jobs. The manned lunar exploration program ended with *Apollo 17* in December 1972. Lunar reflectors left on the moon continue to help us precisely monitor the moon's distance. Study also continues on the 382 kilograms (843 pounds) of lunar rocks and soil returned to Earth by Apollo teams.

In recent years some skeptics have questioned whether or not Americans really set foot on the moon. These critics describe a vast U.S. government conspiracy of lies to fool our Russian competitors in the "space race." However, the idea of a cover-up involving hundreds of thousands of people simply is not credible. Such false stories are hurtful to the families of several dedicated astronauts who perished during the Apollo program. Those who attempt to rewrite and fictionalize history in this way are authors of confusion. We did indeed go to the moon and return safely, decades ago. The footprints and instruments left on the lunar surface are a tribute to the adventurous early years of the U.S. space program.

APOLLO MISSIONS ON THE MOON

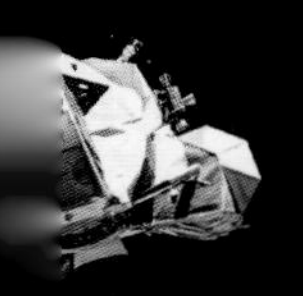

Apollo 15
Landing Site: Hadley Rille/Apennines
David Scott, commander
Alfred Worden, command module pilot
James Irwin, lunar module pilot
Launch Date: July 26, 1971
Moon Landing Date: July 30, 1971
Return Date: August 7, 1971

Apollo 17
Landing Site: Taurus-Littrow
Eugene Cernan, commander
Ronald Evans, command module pilot
Harrison "Jack" Schmitt, lunar module pilot
Launch Date: December 7, 1972
Moon Landing Date: December 11, 1972
Return Date: December 19, 1972

pollo 12
nding Site: Oceanus Procellarum
harles "Pete" Conrad, Jr., commander
chard Gordon, command module pilot
an Bean, lunar module pilot
unch Date: November 14, 1969
oon Landing Date: November 19, 1969
eturn Date: November 24, 1969

Apollo 11
Landing Site: Mare Tranquillitatis
Neil Armstrong, commander
Michael Collins, command module pilot
Edwin "Buzz" Aldrin, Jr., lunar module pilot
Launch Date: July 16,1969
Moon Landing Date: July 20, 1969
Return Date: July 24,1969

Apollo 14
Landing Site: Fra Mauro
Alan Shepard, commander
Stuart Roosa, command module pilot
Edgar Mitchell, lunar module pilot
Launch Date: January 31, 1971
Moon Landing Date: February 5, 1971
Return Date: February 9, 1971

Apollo 16
Landing Site: Descartes
John Young, commander
Thomas Mattingly II, command module pilot
Charles Duke, Jr., lunar module pilot
Launch Date: April 16, 1972
Moon Landing Date: April 21, 1972
Return Date: April 27, 1972

Figure 1.9 Locations of the six Apollo landing sites on the moon with mission dates and emblems

Planet	Number of Moons	Some Moons' Names
Mercury	0	
Venus	0	
Earth	1	
Mars	2	Phobos, Deimos
Jupiter	62	Io, Europa, Ganymede, Callisto
Saturn	33	Mimas, Tethys, Titan
Uranus	27	Ariel, Miranda
Neptune	13	Triton, Nereid

Table 1-3. A list of the known solar system moons that orbit the planets. Selected names are also given.

16. How do other moons compare with ours?

Just as with the eight diverse planets of the solar system, there is great variety in their accompanying moons. Mercury and Venus do not have orbiting moons. Mars has two tiny moons shaped somewhat like potatoes. They have been given the names *Phobos* and *Deimos*, Greek words for fear and terror. Mars, with its reddish color, has long been associated with war and violence.

Jupiter's four largest moons were discovered by Galileo in 1610. *Io* (eye-oh) is brightly colored with red, orange, and yellow shades. Several active volcanoes pour out molten sulfur onto the moon's surface. *Europa* looks like an egg covered with cracks. Its light-colored surface consists of a mixture of frozen chemicals. Craters are lacking, probably obliterated by glacier-like movement of the surface ices. Some astronomers are hopeful that liquid water (and life) exists beneath the icy surface. *Ganymede* is covered with frozen chemicals and craters. It is the largest moon in the solar system with a diameter greater than that of Mercury and Pluto. One giant crater on *Callisto* is a thousand miles in diameter. This unusual feature looks like a bull's-eye target with ten outer rings of circular mountain ranges.

Saturn's *Titan* is the second largest moon in the solar system, nearly half the size of Earth. This moon has surface clouds and an atmosphere of nitrogen gas. It also may have frigid lakes of liquid nitrogen at a temperature below -321°F (-196°C).

The various solar system planets and moons are not closely similar to each other in either color or surface features as a common spontaneous origin would predict. Instead, each space object uniquely shows God's creative variety.

DO YOU KNOW NASA?

- The acronym NASA stands for National Aeronautics and Space Administration.
- NASA started their operations on October 1, 1958, a year after *Sputnik 1* was launched by the Soviet Union.
- The National Advisory Committee for Aeronautics (NACA) was initiated by President Woodrow Wilson to study problems of flight and possible solutions, as well as other flight-related research. NACA was the precursor to NASA.
- The first piloted Apollo mission was *Apollo 7*, which launched October 11, 1968. The astronauts on board were Walter Cunningham, Donn Eisele, and Wally Schirra.
- The command module for the *Apollo 10* was named “Charlie Brown” and the lunar module was named “Snoopy,” from the *Peanuts* comic strip characters.
- *Apollo 11* astronauts ate one of two meals. They ate either bacon squares, peaches, sugar cookie cubes, coffee, and pineapple-grapefruit drink or beef stew, cream of chicken soup, fruitcake, grape punch, and orange drink.
- Armalcolite is a mineral discovered by the *Apollo 11* astronauts and is named for the three involved in that moon mission: ARMstrong, ALdrin, and COLins.
- The *Apollo 10* craft carried 25 U.S. flags to the moon, as well as one 4" x 6" flag for each of the 50 U.S. states.
- All the spacecraft from the Mercury, Gemini, and Apollo missions landed in either the Atlantic or Pacific Oceans when they returned to Earth.

17. Have they recently discovered water on the moon?

There are ongoing efforts to locate water on the moon. The hope is not for the presence of lunar life. That idea was given up decades ago when Apollo astronauts found a sterile, lifeless lunar surface. Instead, a water source might sustain a future lunar base and also provide hydrogen fuel for space missions launched from the moon. Twice, NASA has purposely crashed probes into the lunar surface. The resulting plume of dust and gas was then analyzed for the telltale spectrum, or signature, of water molecules. The 2009 impact site was a crater located near the moon's South Pole. Some experts had suggested that frozen water might have accumulated deep within such craters where the sun never shines and intense cold persists.

The results of the search for water on the moon have not been encouraging. There are indeed slight traces of H_20 on the moon, but not in a recoverable amount. This appears to be the general rule everywhere in space. Water, so essential to life and covering three-quarters of the earth, is lacking elsewhere. In addition, the earth is the only known location in the universe where water readily occurs in all three states of solid, liquid, and gas. Other space locations are extremes of either cold or heat. Meanwhile, each variety of water plays essential roles in the earth's climates and sustainability.

One chief lesson from the space age is the unique nature of planet Earth. As Isaiah 45:18 explains, God did not create the earth to be empty but "formed it to be inhabited." There is truly no other place like home in the known universe.

The LCROSS (Lunar Crater Observation and Sensing Satellite) mission was to search for water on the moon. They did this by sending a rocket crashing into the moon, causing a big impact and creating a crater, throwing tons of debris and, potentially, water, ice, and vapor above the lunar surface. This impact released materials from the lunar surface that were analyzed for the presence of hydrated minerals and told researchers that water was present in trace amounts.

Artist's rendering of the LCROSS spacecraft and Centaur separation. Credit: NASA

Figure 1-10. This image of the moon's south polar region shows a number of potential LCROSS targets. The crater labeled "SP_C" is Cabeus Proper, the final selection. Credit: NASA

Apollo 13 Mission

This mission was intended to land on the moon, but a mid-mission technical malfunction forced the lunar landing to be aborted, the crew returned safely to Earth, and the mission was termed a "successful failure."

MOON PHASE CALENDAR ACTIVITY

Objective:
To chart the different moon phases and the location of the moon in the sky on a blank calendar page, giving the understanding of how the moon phases and location of the moon change daily. For this activity, a blank calendar page and a pencil or pen is all that is needed.

Time:
A 30- or 31-day calendar page will contain a complete 29-day moon cycle (except for a non-leap year February calendar page).

Instructions:
Draw an upside-down T on the first date on the calendar and label it from east to west. At the same time every night during the month, go outside to the same spot facing south, with shoulders pointing east to west. Draw on the calendar page the shape of the moon and where it is in the sky, in relation to the upside-down T pointing east and west. The moon will move and change slightly each night. On nights where it is not visible, or it is cloudy or rainy, draw a cloud or write "not visible."

Learning the movement and phases of the moon:
Each of the four major moon phases (new, first quarter, full moon, and third quarter) occurs about a week apart. A full moon occurs every 29½ days. In between the four listed moon phases, crescent and gibbous moon phases occur. Try to locate the young crescent moon which occurs 2-3 days after the new moon. This young, delicate moon is found in the western sky as the sun sets.

A Trip to the Moon

If you drive to the moon at 70 mph it will take 135 days. Going in a rocket takes about 60 to 70 hours.

CHAPTER WORD REVIEW

breccias — rocks composed of small rock fragments, glass, and soil that have been compacted into cohesive rocks

cold traps — protected spots near the moon's poles that are in permanent shadows

libration — the "rocking" slightly back and forth of the moon in orbit

lowlands — large, flat areas that cover one-half of the moon's visible side

lunar eclipse — occurs when the earth is lined up exactly between the sun and moon

lunar highlands — rugged mountain ranges that appear from Earth as light-colored patches

moon — any natural satellite that orbits a planet

neap tides — during the first and third quarter moon phases, the sun and moon are 90 degrees apart and Earth tides then are lower

regolith — the general lunar surface collection of dust, pebbles, and boulders

sidereal period — the moon's rotation time with respect to the stars (approximately 27½ days)

solar eclipse — when the new moon moves exactly between the earth and sun

spring tides — tides are highest during the full moon and new moon phases when there is a lineup between the sun, earth, and moon and the gravity of both objects then contribute to the tidal pull on the earth

synodic period — the time from one full moon to the next (approximately 29½ days)

"*We leave as we came and, God willing, as we shall return, with peace, and hope for all mankind.*"
Gene Cernan (Apollo 17 commander)

2

HISTORY OF THE MOON

Learning Objectives

1. Compare and contrast some of the older moon origin theories.
2. Relate the most dominant lunar origin theory today.
3. Define the Roche limit.
4. Discuss the moon's assumed evolutionary history.
5. Differentiate the creation view of the moon.
6. Explain why the moon was created on the fourth day and if these days of creation are to be considered literal.
7. Estimate the age of the moon and relate what radioisotope dating tells us about the moon.
8. Discuss whether or not moon dust is an age indicator.
9. Define the terms lunar recession and transient lunar phenomena.
10. Appraise the evolutionary future of the moon.

Words to Recognize

accretion, ex nihilo, plantesimal, radioisotope dating method, Roche limit, yom

1. What are some older moon origin theories?

There is an obvious trend noticeable with origin theories, whether one considers the beginning of life, the moon, or the entire universe. This trend is an unending modification of ideas. Origin theories "roll on by," each with a limited lifetime. Consider some of the theories for the origin of life on Earth: life either began in primordial oceans, or was delivered by a passing comet, or maybe life molecules formed on the surface of clay minerals. Consider also the series of explanations for the entire universe: the steady state theory, the big-bang theory, the inflationary big bang, the plasma theory, interactions with extra dimensions, etc. Perhaps we should not be surprised at the proliferation of origin ideas. If the beginning of all things was supernatural, as Genesis clearly states, then ultimate origins are beyond scientific explanation. By definition, then, any attempt to explain naturally the ultimate origin of *anything* is doomed to failure.

In pre-Apollo days there were three popular naturalistic lunar origin theories. These theories assumed a moon origin billions of years ago by (1) fission or splitting off of material from Earth; (2) capture of an external moon by the earth's gravity; or (3) condensation from nebulous gas and dust at the same time the earth was formed. The following paragraphs further describe each theory.

(1) The Fission Theory

George Darwin (1845–1912), one of Charles Darwin's ten children, championed this view. An expert on the earth's tides, he suggested in 1879 that the moon had originally split off from the earth (Figure 2-1, top). The idea is variously called the *daughter*, *fission*, or *breakaway* theory. George Darwin proposed that the early earth was molten and spinning rapidly. A growing tidal bulge formed due to the sun and then separated or fissioned from the earth, somewhat similar to mud thrown from a bicycle wheel. The Pacific Ocean basin was later suggested as the "scar" left behind by the lost lunar material.

There are at least four major problems with the fission theory. First, it requires a rotation time for the earth nearly ten times faster than at present, 2.6 hours compared with 24 hours. Historical evidence for such a rapid rotation is lacking. And what became of this large turning motion, called *angular momentum*? Such motion is a constant or *conserved* quantity, and the combined earth-moon system today has only half of the turning motion necessary for original rotational instability and fission. A second problem is that moon materials are not a close match with the earth's composition. For example, iron makes up only about 2 percent of the moon's mass, but more than 30 percent of the earth's. In addition, there is not a significant amount of water on the moon. Third, the moon's orbit is inclined to the earth's equator by an angle that varies between 18½° and 28½°. A fission process would have inserted the moon into an orbit closely aligned with Earth's equator. As a fourth serious problem, a moon thrown off from the earth must pass through the *Roche limit*. This is a distance range in the vicinity of any planet where a moon will be broken up by strong gravity forces. For example, Saturn's rings lie within the planet's Roche limit. A moon passing through this limit close to the earth would likewise be shattered into small fragments, forming a ring around the earth.

(2) The Capture Theory

In this theory, a wandering moon long ago approached the earth and was trapped by the planet's strong gravity. Since, according to this theory, the moon came from somewhere else in the solar system, this is not really an *origin* theory for the moon. A second problem involves the requirement to somehow slow down the wandering moon and insert it into permanent Earth orbit. However, there is no known means by which the moon's velocity could be largely dissipated on a single pass. No other moons, comets, or asteroids have been observed changing into planetary satellites by such a capture process. A third basic problem concerns the moon's highly circular orbit. Capture, if somehow possible, would more likely lead to a greatly elongated, elliptical orbit.

As with fission, lunar capture is recognized today by most astronomers as a doubtful explanation. Of course, even if major physical problems did not exist, this would not demand that the moon formed naturally by any of these natural mechanisms.

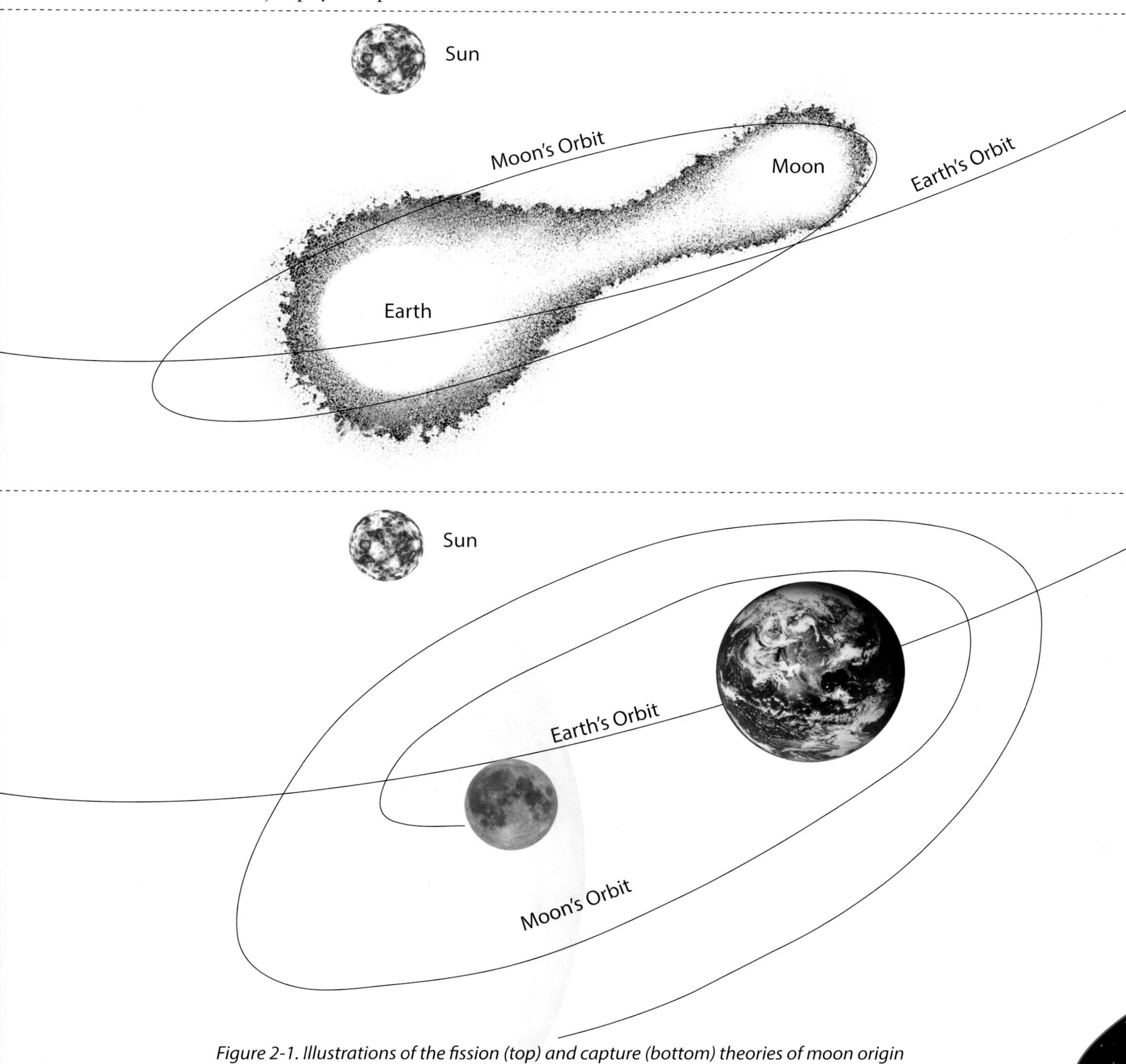

Figure 2-1. Illustrations of the fission (top) and capture (bottom) theories of moon origin

(3) The Nebular Theory

The *nebular*, *sister*, or *condensation* theory calls for an *accretion* or growth of the earth and its moon from dust and gas occurring side by side in space (Figure 2-2). One crucial problem is the precise balance needed during this proposed build-up of solid matter. As the objects grow in size, both the gravity attraction and the moon's orbital speed must increase very precisely to avoid lunar escape or a collision with the earth.

A major assumption here is that nebular contraction will occur in the first place. In general, nebular clouds in space are observed to expand and dissipate due to outward gas forces. To become stable, such clouds need to be compressed externally until inward gravity finally becomes the dominant force. Such is the case for the sun and the outer gaseous planets. For our moon, the details have not yet been worked out by computer models for an external compression and gravity collapse of gas to form a solid object. Typical gas clouds observed in space are thousands of times larger than the moon, and are expanding still further due to outward forces.

In considering the weaknesses of the naturalistic lunar-origin hypotheses, it is good to realize that only a relatively few individuals originate and defend these broad theories. Although origin views receive high priority in the news media, the vast majority of scientists are not strongly involved in the debate over such ideas. In fact, a subject such as lunar origins simply is not representative of scientific inquiry since it is not reproducible. As one Apollo engineer explained it:

> You've got to realize that we've lived with some of these theories for so long that they don't mean much to us anymore. To a large extent, these are matters of pure speculation. We've heard the same old people spin out the same old cobwebs and speculations for years without adding much to them (Mitroff, 1974, p. 60).

When all other ideas are seen to fail, the world is faced with the full power and truth of the Genesis record of the creation of the moon and the universe beyond. It is the one proposition that fits all known facts. Unfortunately, it is also the one proposition that is almost universally scorned and suppressed in spite of the fact that those who do so are empty-handed.

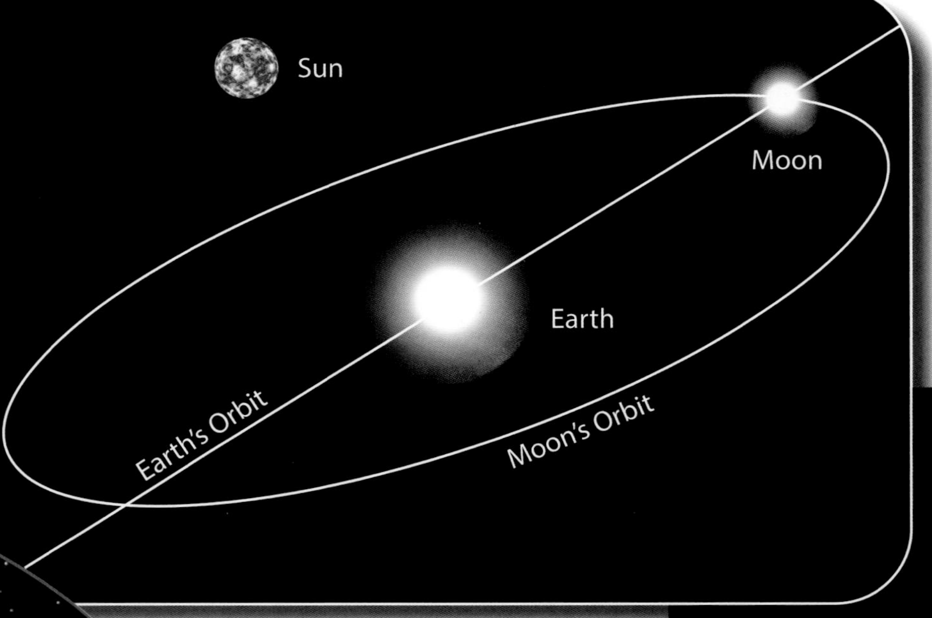

Figure 2-2. Illustration of the nebular (left) and collision (bottom) theories of moon origin

2. What lunar origin theory dominates today?

Serious problems exist for the traditional lunar origin theories discussed in the previous question. As a result, in recent years a new idea has arisen called the *collision* or *giant-impact* theory. In this scenario, the early earth was hit in a chance collision about 4.5 billion years ago by a Mars-sized object traveling through space at 25,000 mph. The great collision partially melted both objects and threw massive debris into Earth's orbit. The earth then slowly recovered its spherical shape and the space debris finally coalesced into the orbiting moon.

The lunar collision theory is currently popular because it avoids some weaknesses of the previous origin theories. For example, the moon is less dense than the earth and apparently lacks an iron core. The suggestion is that the ejected debris came from the earth's iron-deficient outer layers, leaving the iron behind, which became concentrated into the earth's core.

It should be remembered that this story of lunar origin by collision is only a model and not reality. Scientists perform computer simulations with many variables, including the colliding object's size, speed, composition, and angle of impact. Some early moon collision models require two separate collisions with Earth; other models result in multiple moons orbiting the earth.

Like the nebular theory, lunar collision faces the problem of how a ring of debris actually will draw together into a moon. The earth's Roche limit remains a serious physical problem for moon formation by collision (Lissauer, 1997). Added to these problems is the low probability of a collision between planet-sized objects in the first place. And there is more than one moon to explain; there are over 100 solar system moons. It is very difficult to imagine all these space objects being formed by improbable collision events.

3. What is the Roche limit?

French astronomer Edouard Roche studied gravity effects on satellite moons. In 1849 he showed mathematically that a moon would be broken into pieces if it was located close to its host planet. This breakup happens because the planet pulls much more strongly on the near side of the moon than its far side. This force difference may exceed the moon's own internal gravity force. Breakup occurs within about 2.44 planetary radii, a distance now called the *Roche limit.*

The rings of Saturn are closer to the planet than the distance at which a large solid body can exist — that is, they are within Saturn's Roche limit. For the earth-moon combination, the breakup distance is about 11,500 miles (18,500 km) from the earth's center. Our distant moon is stable, being 21 times as far away as the breakup instance. The stability limit does not apply to nearby man-made satellites and space probes that are held together with welds and rivets instead of gravity alone.

There are several implications involving the Roche limit. For example, it presents a serious problem for the fission and collision theories of lunar origin. Both theories involve a moon initially positioned in the vicinity of Earth. In this case, we would end up with a Saturn-like ring of debris instead of a solid moon.

The distance limit for a stable moon is also a basic problem for evolutionary age assumptions. Due to tidal effects, the moon is very slowly spiraling outward from the earth. The average earth-moon separation is presently increasing by 1 to 2 inches each year. Extrapolating backward in time, the moon was closer to us last year, and closer yet millennia ago. Analysis shows that the moon would have been in close contact with the earth about 1.4 billion years ago. This is not to say the moon is this old, but that the moon *cannot* possibly be this old or it would have disintegrated into fragments. However, the moon is typically said to be 4.6 billion years old. The Roche limit raises a fundamental challenge to this ancient age assumption.

4. What is the moon's assumed evolutionary history?

A long sequence of historical events has been constructed to explain the appearance of the moon. The story actually begins 10–15 billion years ago with the *big-bang* theory. As the early universe expanded, clouds of hydrogen and helium formed initially. These clouds later collapsed into stars. Within the cores of these ancient stars, the heavier elements (carbon, iron, etc.) were made by nuclear fusion reactions. When these early stars eventually disintegrated, the elements were then spread across space. In our vicinity of the universe, some of the stellar debris eventually came together to begin the early solar system. Many large objects are said to have resulted from the *accretion* or gathering together of dust and gas.

The moon is said to have been born about 4.6 billion years ago by the collision of a large Mars-sized space object, called a *planetesimal*, with the early earth. After cooling from a molten state, the moon's hardened surface was pummeled by space debris, which formed many craters. During the next billion years, the moon's interior melted due to the decay of radioactive elements. Internal molten rock then flooded upward to form the maria basins or seas of basalt. The lava flows gradually ceased about 3 billion years ago, and the number of falling meteorites gradually diminished. As a result, the seas themselves are only lightly cratered. Since then, the moon has remained inactive with little change to the present day. The moon is therefore described as an unchanging "museum" of the early solar system.

5. What is the creation view of the moon?

To the Hebrew mind, the concept of the created moon, together with the sun and stars, served to magnify the glory of God. This follows because of the special way in which the opening chapter of the Bible sets forth this cosmic event. In total contrast to the wide spectrum of creation concepts that characterizes ancient paganism and modern naturalism, the traditional Hebrew/Christian understanding of the opening chapter of Genesis has been simple and straightforward. At least two things can be clearly discerned in the creation record that unveil the absolute glory of the Creator. First, the astronomical bodies were created suddenly, thus establishing the overwhelming uniqueness, in fact, the absolute ultimacy of God's power or omnipotence. Second, the astronomical bodies were created after the earth and plant life had already been created. This eliminates all potential competition for the claim of final sovereignty and deity. This includes not only solar or lunar deities, but also the modern secular "god" of cosmic evolution. The first of these two profoundly important facts of special revelation, concerning creation's timescale, will now be analyzed in more detail.

The creation of the universe was not only *ex nihilo* (i.e., from no previously existing matter, as stated in Heb. 11:3), but it was also, by the very nature of the case, *instantaneous*. Its origin was not, therefore, spontaneous or self-acting. The evolutionary concept of a gradual buildup of heavier and heavier elements throughout billions of years is clearly excluded by the clear statements of Scripture.

The immediate effect of God's creative word is emphatically stated in Psalm 33:6–9: "By the word of the LORD the heavens were made, and all the host of them by the breath of his mouth. . . . For He spoke, and it was done; He commanded, and it stood fast." There is certainly no thought here of gradual development, or trial and error process, or age-long, step-by-step fulfillment. In fact, it is quite impossible to imagine any time interval in the transition from absolute nonexistence to existence! Similarly, "God said, 'Let there be light'; and there was light" (Gen. 1:3). At one moment, there was no light anywhere in the universe; the next moment, physical light existed in abundance. So spectacular is this creation event that the New Testament compares it to the suddenness and supernaturalness of conversion (2 Cor. 4:4–6; 5:17). It may be confidently asserted that the idea of *sudden appearance* dominates the entire creation account (see Gen. 1:1, 3, 12, 16, 21, 25, 27; 2:7, 19, 22).

One is confronted, however, with numerous contemporary denials of this concept. It is frequently claimed, for example, that God's command to the earth to "bring forth grass, the herb that yields seed, and the fruit tree that yields pgs 16 -

- water on the moon
  1% 2% water
  - cold traps - protected spots near moon's poles (possibly in craters) in permanent shadows that may have vast deposits of water (ice)
  - Lunar probes Clementine + Lunar Prospector (1999) did not find any.
  - ? water delivered by meteorites or comets in distant past ?

- Fear of microbes from the moon dessimating life on Earth
  - quarantine dropped after 3rd Apollo landing

- ~~No~~ lack of carbon, ~~no~~ lack of free Oxygen (traces found was possibly contaminants)

- life on other planets / moons
  - spontaneous evolution (abiogenesis)
  - supernatural creation
  - contamination from Earth

- Life on Earth is unique Isaiah 45:18
- Earth microbes have inevitably gone to moon, Venus, Mars, Jupiter, etc as passengers on spacecraft

pg 17 phases of the moon

- Moon phases affect tides
  - Tides are highest during full moon + new moon

pg 18

gravity from sun, Earth, moon line-up contributes to tidal pull. Called "spring tides" (not seasonal)

≈ 2/3 due to moon; 1/3 sun ← gravity

@ 1st + 3rd qtr, sun + moon @ 90° ∡, low tides neap tides.

[illegible] the [illegible]

[illegible] [illegible]

[illegible] dated [illegible]

[illegible] called [illegible]

that may have used deposits [illegible]

[illegible]

[illegible] [illegible]

[illegible]

[illegible] the plants/[illegible]

[illegible] (changed [illegible])

[illegible]

[illegible]

[illegible] have [illegible] complete [illegible]

[illegible] [illegible]

[illegible]

[illegible] [illegible]

according to its kind" (Gen. 1:11) implies a long process under the providential direction of God. But Numbers 17:8 is the true analogy to Genesis 1:11, for in one night Aaron's rod "sprouted and put forth buds and produced blossoms ... and yielded ripe almonds." (See also Jonah 4:6–10.) It is significant that Russell Maatman, a proponent of the day-age theory, admits that "there is no doubt that *each creation event* was instantaneous. One moment a certain thing existed; the previous moment, it did not exist" (*The Bible, Natural Science and Evolution*, 1970, p. 95). It must be made clear that theologians who emphasize the suddenness of God's *creative* acts and sign miracles do not thereby minimize the glory of God's non-miraculous *providential* works in human history (see Dan. 4:17 and the Book of Esther). Miracles and providence are *not* identical and dare not be confused. Thus, the conception of Jesus was sudden and supernatural, while His birth was the result of a gradual and natural process carried out under the providential control of God. This distinction is profoundly important, for if the conception of Jesus is understood to be only providential but not miraculous, the incarnation is denied and Christianity is destroyed (see 1 John 4:3; 2 John 7). Likewise, if the events of Genesis 1–2 are understood to be providential but not miraculous, creationism is not simply modified; it also is destroyed.

This leads us to a second important consideration pertaining to the sudden creation of the astronomical universe, namely, the analogy of God's creative works in the person of Christ during His earthly ministry two thousand years ago in the Holy Land. Since the New Testament makes it clear that the universe was created through Christ, the Son of God (John 1:3, 10; Col. 1:16; Heb. 1:2), and that the miracles He performed while on Earth were intended to reveal His true nature and glory (John 1:14, 2:11, 20:31), it is deeply instructive to note that these works all involved sudden transformations. Thus, while one philosopher has claimed that there is "no strategy as slippery and dangerous as analogy," the biblical analogy of Christ's creative work in Genesis and in the Gospels remains irresistibly powerful.

In response to the mere word of Jesus Christ, for example, a raging storm *suddenly* ceased, a large supply of food *suddenly* came into existence, a man born blind *suddenly* had his sight restored, and a dead man *suddenly* stood at the entrance of his tomb. Of the vast number of healing miracles performed by Christ, the only recorded exception to instantaneous cures is that of the blind man whose sight was restored in two stages, each stage, however, being instantaneous (Mark 8:22–26). Such miracles were undeniable signs of supernaturalism in our Lord's public claim to messiahship, and we may be quite sure that if, in His healing of the sick and crippled and blind, He had exhibited "the prodigal disregard for the passing of time that marks the hand of him who fashions a work of art," no one would have paid any attention to His claims! If the Sea of Galilee had required two days to calm down after Jesus said, "Peace, be still," the disciples would neither have "feared exceedingly," nor would they have "said to one another, 'Who can this be, that even the wind and the sea obey Him?'" (Mark 4:39–41). Every effort to modify the suddenness and supernaturalness of creation events to make them more acceptable to "the modern mind" only results in the long run in minimizing and obscuring the true attributes of the God of creation. This has been a difficult lesson for many Christians to learn.

6. Why was the moon created on the fourth day?

The order of events in Genesis 1 is deliberate and meaningful. At the same time, however, the creation sequence is very confusing to the natural mind. How could the earth (day 1) be created before the sun (day 4)? What did the earth orbit during the first three days, and what was the original light source? How could liquid water (day 1) and plants (day 3) possibly exist before the sun? Would not everything freeze solid instantly? How could whales (day 5) appear before the land mammals (day 6) from which whales supposedly evolved? Such questions continue without end when the natural mind attempts to understand the details of the creation week. However, the creation events are supernatural from start to finish and are therefore *holy ground*. Creation details lie forever beyond scientific explanation.

The creation of the sun, moon, and stars on the fourth day of creation teaches a profound lesson: The Creator is infinitely superior to His works, including the heavens. Therefore, no visible heavenly body, including the moon, should be worshiped. This profound theological principle is confirmed and demonstrated biblically by the fact that the moon was directly and instantaneously brought into existence apart from pre-existent materials by the spoken word of the transcendent God of the universe. Therefore, He alone is to be worshiped by men. Furthermore — and this neglected fact is theologically crucial — the infinite inferiority of the sun and moon to the true God of creation is fixed by their having been created after the creation of the earth and its vegetation. In the words of a widely read professor of the history of science:

"In the first chapter of Genesis it is made evident that absolutely nothing, except God, has any claim to divinity; even the sun and moon, supreme gods of the neighboring peoples [of Israel], are set in their places between the herbs and the animals and are brought into the service of mankind" (Hooykaas, 1972, p. 8).

7. Were the days of creation literal?

The creation week has been interpreted various ways. In theistic evolution, God providentially directed the cosmos toward higher levels of complexity using macroscopic evolution throughout long ages. In progressive creation and the day-age theory, God occasionally introduced new kinds of plants, animals, and finally man into a geologic timetable spanning billions of years. The gap theory of Genesis 1:1–2 has God creating a perfect world perhaps billions of years ago, destroying it all at the fall of Satan, and then later re-creating it in six literal days.

In total contrast to these views, the traditional Hebrew/Christian understanding of the opening chapter of Genesis is straightforward with six literal days of supernatural creation. There is a very good reason for this interpretation. All other historical narratives in the Bible likewise were understood in a normal manner. This is technically known as the historical/grammatical method of hermeneutics, which takes into full account not only the context of each passage but also the known literary figures of speech. Since there is no evidence of poetry in the first chapter of Genesis, it seems rather obvious that God intended the chapter to be understood normally. If, on the other hand, we abandon this God-honored and time-honored method of historical/grammatical interpretation, then all hope of definitively determining what the opening statements of the Bible really mean must be abandoned.

The fact that God's work of creation was completed in six literal days clearly demonstrates that the creative work of each day was sudden and supernatural. In view of the widespread resistance to this concept, even in some Christian circles, it may be surprising to many people to learn how strong are the biblical arguments in its support. Four of these arguments as well as answers to major objections will be presented here.

(1) Although the Hebrew word for day, "*yom,*" can refer to a time period longer than 24 hours if the context demands it (e.g., "day of the Lord") its attachment to a numerical adjective restricts its meaning to 24 hours ("first day," etc., with a precise parallel in Num. 7:12–78). The expression "one day" in Zechariah 14:7, claimed by some to be an exception, probably refers to a literal day also, because of the term "evening" in the same verse.

(2) The qualifying phrase, "evening and the morning," which is attached to each of the creation days throughout Genesis 1, indicates a 24-hour cycle of the earth rotating on its axis in reference to some fixed astronomical light source (not necessarily the sun). The same phrase appears in Daniel 8:26 (see also Dan. 8:14; ASV, NASB) where it must be understood literally. Some have claimed Psalm 90:6 as an example of a figurative use of the "evening" and the "morning." However, even this example is questionable, for the Genesis 1 formula is not used, and "morning" appears before "evening." Furthermore, even if "morning" and "evening" in Psalm 90:6 are used in a figurative sense, the figure would be meaningless if it did not presuppose the literal use of such terms in earlier historical narratives of Scripture, such as Genesis 1.

(3) A creation week of six indefinite periods of time would hardly serve as a valid and meaningful pattern for Israel's cycle of work and rest, as explained by God in the fourth commandment (Exod. 20:11, 31:17). While it is, of course, true that God *could* have created the world in six microseconds or in six trillion years if He had chosen to do so, such speculation is completely irrelevant in the face of the fourth commandment, which informs us that God, as a matter of fact, chose to create the world in six days in order to provide a clear pattern for Israel's work and rest periods. The phrase "six days" (note the plural) can hardly be figurative in such a context.

(4) Since the word "days" in Genesis 1:14 is linked with the word "years," it is quite obvious that our well-known units of time are being referred to, their duration being determined not by cultural or subjective

circumstances, but by the fixed movements of the earth in reference to the sun. Otherwise, the term "years" would be meaningless. We must assume that the first three days of the creation week were the same length as the last three astronomically fixed days, because exactly the same descriptive phrases are used for each of the six days (e.g., numerical adjectives and the evening/morning formula) and all six days are grouped together in Exodus 20:11 to serve as a model for Israel's work week. The fact that the sun was not created until the fourth day does not make the first three days indefinite periods of time, for on the first day God created a fixed and localized light source in the heavens in reference to which the rotating earth passed through the same day/night cycle. To interpret the word "day" in this chapter as a long or indefinite period of time is thus completely arbitrary. Instead, the universe was created by God within one literal week.

In opposition to the literal-day interpretation, it has been asserted that Arctic people experience a six-month day instead of a 24-hour day. This argument is not valid. Even during winter months, polar inhabitants can observe from the alternating phases of light on the horizon that a day lasts 24 hours. Pressed to its logical conclusion, such an argument would mean that the word "day" could *never* mean a 24-hour period, either in the Bible or in our contemporary experience.

It has also been maintained that other passages of the Bible speak of a day in God's sight being as a thousand years. The Bible indeed makes this statement (Ps. 90:4; 2 Pet. 3:8); but instead of destroying the literal-day interpretation of Genesis 1, it actually helps to establish it. In 2 Peter 3:8, for example, we are *not* told that God's days last a thousand years each, but that "with the Lord one day is *as* a thousand years." To say "*as* a thousand years" is a very different matter from saying "*is* a thousand years." This point has often been overlooked. If "one day" in this verse really means a long period of time, then we would have the following absurdity: "a long period of time is with the Lord as a thousand years."

But a thousand years would be a long period of time for human beings, too! It also should be realized that an "old earth" requires the days of creation to be much longer than 1,000 years each. In the day age theory, for example, each creation day is nearly a billion years long.

The obvious teaching of 2 Peter 3:8 (as well as Ps. 90:4), then, is that God exists above the limitations of time. One valid deduction from this fact is that God can accomplish in one brief, literal day what man could not accomplish in a thousand years, if ever. This is one of the astounding messages that comes through to us from the creation narrative of Genesis 1: God has infinite power; we do not! The prophet Jeremiah understood this: "Ah, Lord God! Behold, You have made the heavens and the earth by Your great power and outstretched arm. There is nothing too hard for You" (Jer. 32:17). Did the Creator, perhaps, really need the seventh day to rest from six days of creative work? The answer comes back with overwhelming clarity: "Have you not known? Have you not heard? The everlasting God, the LORD, the Creator of the ends of the earth, neither faints nor is weary. . . . He gives power to the weak; and to those who have no might He increases strength" (Isa. 40:28–29).

There is, in fact, simply no way for the human mind to grasp the power of God:

> *"To whom then will you liken Me, or to whom shall I be equal?" says the Holy One. . . . "For as the heavens are higher than the earth, so are My ways higher than your ways, and My thoughts than your thoughts" (Isa. 40:25; 55:9).*

This enormously significant truth concerning God is shattered when one tries to stretch the account of creation to incorporate vast ages of time in order to make the passage more "reasonable," and thus to accommodate it to man's finite, uniformitarian level of thinking. To twist the Scripture in this way is to distort God's message to us. What the Apostle Peter said concerning Paul's letters is therefore completely applicable to the opening chapters of the Bible: "In which are some things hard to understand, which untaught and unstable people twist to their own destruction, as they do also the rest of the Scriptures" (2 Pet. 3:16).

Another widely held objection to the literal-day interpretation of Genesis 1 is that the seventh day never terminated because God is *still* resting from His work of creation (Heb. 4:3–11). But this argument introduces much confusion between historical events and their spiritual application. The "rest" of Hebrews 4 is primarily the spiritual rest of salvation (Matt. 11:28–30) whereby the believer shares in the eternal blessedness and fulfillment that characterizes God. Certainly God did not have to wait until the end of the sixth day of creation week for this kind of rest to begin!

Thus, the first Sabbath was not instituted for God's benefit (Mark 2:27). It is this often neglected point that is crucial in determining the duration of the original Sabbath day.

How long, then, did the first Sabbath continue? It is obvious that all Israelites, to whom Sabbath observance was specifically applied by God, understood this period to be exactly 24 hours in length, based on the pattern of God's creation Sabbath:

> *Six days you shall labor, and do all your work, but the seventh day is the Sabbath of the LORD your God. In it you shall do no work. . . . For in six days the LORD made the heavens and the earth, the sea, and all that is in them, and rested the seventh day. Therefore the LORD blessed the Sabbath day and hallowed it* (Exod. 20:9–11).

Any Israelite who decided to extend his Sabbath observance indefinitely on the assumption that God's Sabbath still continues would have starved to death (see Exod. 35:3). Equally significant is the deduction that Adam and Eve must have lived through the entire seventh day of creation week before God drove them out of the garden, for God would not have cursed the ground (Gen. 3:17) during the very day He "blessed" and "sanctified" (Gen. 2:3).

In conclusion, there is simply no escaping the fact that God intends us to understand the creation of the astronomical bodies, including the moon, to have been instantaneous. The implications of this profound fact with regard to currently popular attempts to harmonize Genesis with cosmic evolution should be perfectly obvious. To suggest a "gradual creation" of the moon may be conceivable to some minds. But for most people, such a concept would raise the very serious question as to whether God, as a matter of fact, ever created the moon at all. When the stupendous fact begins to dawn upon us, however, that the moon was created *instantaneously* and *ex nihilo*, all serious questions concerning the deity, omnipotence, and glory of the moon's Creator evaporate. This is why the Hebrew/Christian approach to creation is shocking and transforming in its impact upon the human mind.

8. How old is the moon?

The standard secular age assigned to the moon is about 4.6 billion years. This same age is also applied to the sun, the earth, and other planets of the solar system. It is derived from the radioisotope study of moon rocks and also meteorites. However, no scientific method can date a rock with absolute certainty. The *young earth* creation view supports a much smaller value for the age of the moon, solar system, and universe. The moon is thought to be between 6,000 and 10,000 years old. This number is at least a half-million times smaller than the evolutionary 4.6 billion years. The creationist age is based upon two factors. *First*, there is a healthy skepticism of radioisotope dating results. Many hundreds of scientists worldwide question the vast ages assigned to rocks. These researchers are very interested in overlooked kinds of science data that support a recent creation. The *second* factor supporting a recent creation is the record of biblical history. The Bible most naturally allows for thousands of years of history, but not millions or billions of years.

RADIOISOTOPE METHOD

Parent Isotope		Daughter Isotope		Half-life (billions of years)
^{14}C	Carbon	^{14}N	Nitrogen	.000005730 (5730 years)
^{40}K	Potassium	^{40}Ar ^{40}Ca	Argon Calcium	1.25
^{87}Rb	Rubidium	^{87}Sr	Strontium	48.8
^{147}Sm	Samarium	^{143}Nd	Neodymium	106
^{176}Lu	Lutetium	^{176}Hf	Hafnium	35
^{187}Re	Rhenium	^{187}Os	Osium	43
^{232}Th	Thorium	^{208}Pb	Lead	14.1
^{235}U	Carbon	^{207}Pb		0.704
^{238}U		^{206}Pb		4.47

Table 2-1. Radioactive isotopes commonly used in dating. The first entry, carbon 14, is applied to once living organisms, including plants and animals. The other isotopes are commonly applied to rock formations.

9. What does radioisotope dating tell us about the moon?

Certain kinds of atoms are unstable or radioactive. This is usually because of an excess number of neutrons in their atomic nucleus. After a period of time the atom disintegrates, giving off radiation in the process. The original "parent" atom changes into an entirely different "daughter" atom. As examples, carbon 14 turns into nitrogen, and uranium eventually turns into the element lead. Now suppose an object is originally formed with a certain number of parent atoms within it. As this object ages, the number of parent atoms decreases and the number of daughter atoms increases. If the "half-life" of the parent atom is known, an age for the object can then be calculated from the later composition of its atoms. Half-life measures the average lifetime of radioactive atoms. During one half-life, 50 percent of a collection of the atoms will decay. Table 2-1 gives some examples of decaying atoms and their half-lives. Except for carbon 14, the listed isotope examples are commonly used for the dating of rocks.

Carbon 14 dating is a special category, not applied to rocks from the earth or moon. It is limited to a much shorter timescale than the other isotopes. The technique involves the absorption of atmospheric carbon 14 by living plants or animals. After their demise, carbon 14 then slowly disintegrates within the preserved organic remains. Isotope measurements reveal the time when the organism was alive. The carbon 14 technique has proven useful for the dating of many archaeological artifacts that incorporate plant material.

The radioisotope dating method involves chemical analysis to determine the ratios of parent and daughter atoms in the sample. For rock samples, these include the ratios of uranium to lead, rubidium to strontium, and potassium to argon. This analysis is very precise, measuring parts per billion or parts per trillion of atoms. Age results in the range of millions or billions of years are not unusual. However, these calculated ages are always subject to interpretation. The number of atoms can be accurately measured, but what does this data really mean? Assumptions must be made about the initial composition of the sample along with any changes that may have occurred during the history of the sample. For example, parent or daughter atoms may have migrated into or out of the rock sample, thus invalidating the calculated age. Also, many creation scientists believe that radioactive decay was *accelerated* at some point in history, perhaps at the time of creation, the Curse, or else during the year of the Genesis flood. Whatever might have triggered the acceleration of nuclear decay, it involved rock samples everywhere: the earth, moon, Mars, etc. The result was a rapid accumulation of daughter isotopes and apparent great age.

Some lunar rocks give remarkably old ages, around 4.6 billion years. Certain meteorites also date this old. However, no Earth rocks have been found with this age; most are at least a billion years younger. The usual conclusion is that Earth rocks have been

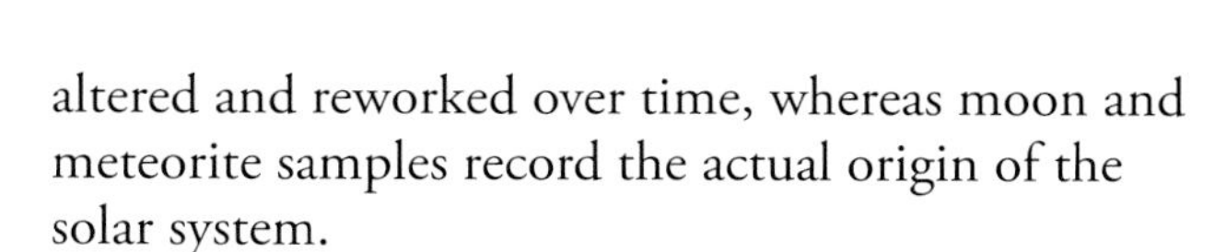

altered and reworked over time, whereas moon and meteorite samples record the actual origin of the solar system.

Creationists appear to be the only group that challenges radioisotope dating results. And the assumption of an ancient earth and universe indeed needs to be questioned. Radioisotope dating should be placed in a similar category to other scientific dating techniques. All of them are fallible and none should be solely relied upon, whether they give an ancient age or a recent age for the earth or moon. Unfortunately, the interpretation of radiometric dating results is intertwined with the evolutionary assumption of an old universe. The reinterpretation of such results is an active area of creation research.

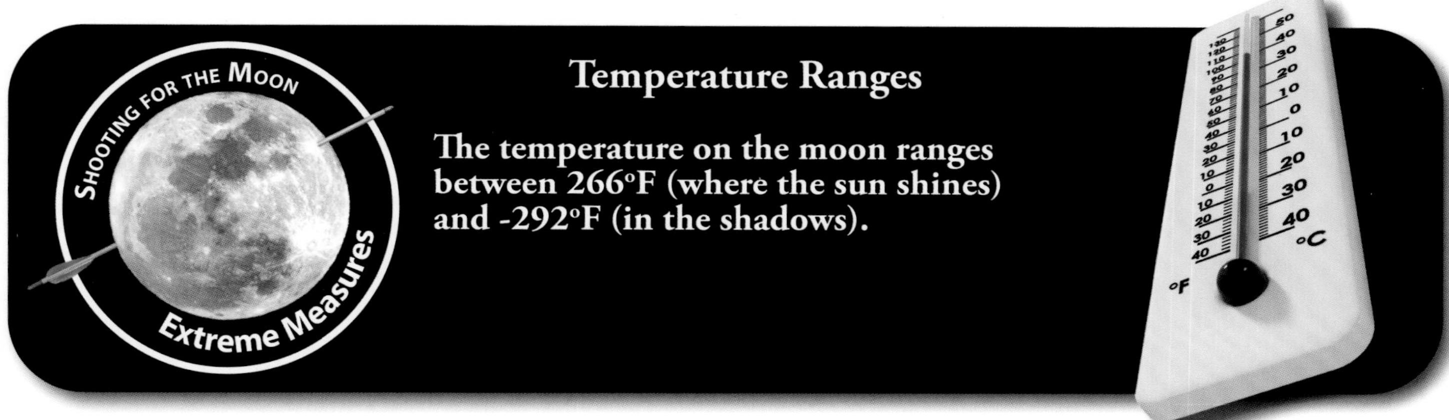

Shooting for the Moon
Extreme Measures
Temperature Ranges
The temperature on the moon ranges between 266°F (where the sun shines) and -292°F (in the shadows).
°F
°C

10. Is moon dust an age indicator?

Both the earth and moon collect dust that continually rains downward from space. On the earth, these dust particles are rapidly swept away by wind and water. On the moon, however, there are no weather elements to move the dust around. The moon therefore collects an ever-growing layer of dust and meteorite fragments.

The resulting depth of lunar dust layers should provide an age indicator. However, a major uncertainty involves the rate of dust fall or *influx*, both in the past and present. The difficult measurement is typically done by high-altitude aircraft that collect stray dust. In the pre-Apollo years there were some estimates of a high rate of dust accumulation, resulting in a dangerously deep layer of dust on the moon over a multi-billion-year time span. In this case, lunar astronauts might stir up a great cloud of dust and become disoriented, if not buried completely. Before the manned missions, several Surveyor spacecraft were safely landed on the moon. They revealed only a thin dust layer. More recent influx measurements also give a small rate, in apparent agreement with the 2–3-inch-thick layer found on the moon, even over a long time span.

Two creationist responses are in order. *First*, the evolutionary view predicts a much greater amount of space dust and debris in the early stages of solar system formation. On this basis, a thick layer of existing moon dust still would be expected on a long time scale. *Second*, the numerical value of the rate of dust influx remains quite uncertain. Creationists await further measurements of dust falls on the earth and moon. Meanwhile, it remains an open question whether or not lunar dust is a strong argument for a recent creation.

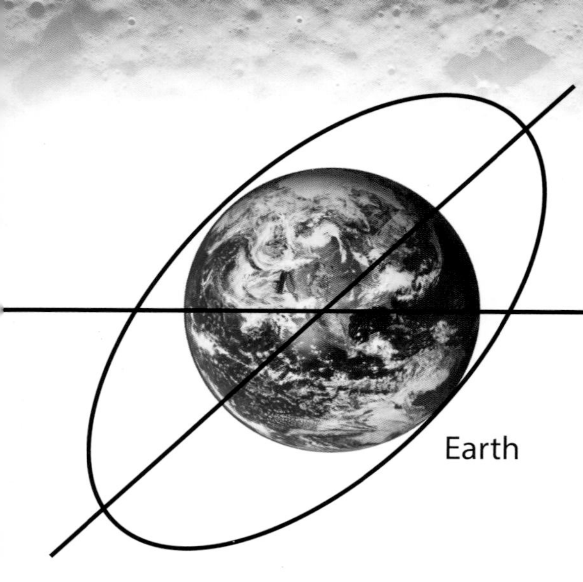

Figure 2-3. The moon's gravity attraction results in tidal bulges on the earth, about ten degrees east of the moon's location. The angle is exaggerated in the figure for clarity. As a result, Earth rotation is gradually slowing and the earth-moon separation is increasing.

11. What is lunar recession?

The moon is slowly spiraling outward from the earth. The earth-moon separation distance currently is increasing by about 4 cm (1.6 in) each year. This effect is due to the gravity interaction between the earth and moon. The earth's tidal bulges actually occur somewhat east of the moon's location. The earth's rapid rotation carries the tidal bulge about ten degrees ahead of a line connecting the earth and moon (Figure 2-3). The result is a continuous, slight forward tug on the moon, causing it to gradually lift into an ever-larger orbit around the earth. There is also a slight braking effect on the earth's spin due to friction of ocean movement. This slows the earth's rotation by about two milliseconds each century.

The rate of separation, very small today, is strongly dependent on the total earth-moon separation. This rate varies as the sixth power of the actual separation (DeYoung, 1990). This means that in earlier times the moon would have moved outward much more rapidly than is measured today. Extrapolation backward in time shows that the moon would have been in the near vicinity of the earth about 1.4 billion years ago. However, this result directly conflicts with the assumed age of the moon, 4.6 billion years. Furthermore, a nearby moon would result in immense tides and probable melting of the earth and moon. The moon also would disintegrate when inside the earth's Roche limit. Clearly, a time-scale problem exists for an ancient moon. In contrast, the earth-moon distance does not change appreciably on a 10,000-year time scale, only about one-half mile. Thus, the recent-creation view avoids the close-approach problem of the earth-moon system.

Moon Golf

Alan Sheppard hit a golf ball while out walking on the moon. It sailed 2,400 feet!

12. What are transient lunar phenomena?

The moon is commonly described as a cold, dead, unchanging satellite of the earth. Geologic activity and impact collisions are assumed to have ceased about three billion years ago. The moon has been called the "museum of the early solar system." However, there is a growing list of observations of changes still occurring on the moon. These are given the name *transient lunar phenomena*, or TLPs. They include local color changes, glowing clouds, bright spots, streaks of light, hazes, and mists. There is also measurable lunar seismic or quake activity, although at a greatly reduced level from the earth. A 1968 publication from the National Aeronautics and Space Administration (NASA), *Chronological Catalog of Reported Lunar Events*, contains 579 separate entries and 250 references. There is also a compilation of TLPs by William Corliss (Corliss, 1975).

By definition, TLPs are short-lived and therefore difficult to observe or verify. The lunar events generally cover an area of only a few kilometers' extent and last a few hours. By the time one can notify a second observer about the TLP, it may have ceased. This transitory nature of TLPs probably discourages the reporting of many lunar sightings because of uncertainty or fear of ridicule.

The TLPs have a rich history. Astronomer William Herschel, during 1783–1787, reported several lunar volcanic emissions. Here is one of Herschel's descriptions (Ley, 1965, p. 71): "I perceived in the dark part of the moon a luminous spot. It had the appearance of a red star. . . . I [also] perceived three volcanoes. . . . The third showed an actual eruption of fire or luminous matter." Many additional TLP events are described in the first edition of this book (Whitcomb and DeYoung, 1978). Dozens of TLPs have been associated with the particular craters Aristarchus, Plato, and Alphonsus. These transients may be signs of gas discharges or volcanic activity. This is totally unexpected if the moon is truly ancient.

It is likely that the TLPs are energized by heat and pressure from the lunar interior. Other minor factors may include a fluorescent effect from solar ultraviolet light and electrostatic lightning. Whatever the mechanisms involved, the transient lunar phenomena are present. All of the reports cannot be ascribed to faulty observing techniques or erroneous interpretations. These transient events indicate that the moon is still active geologically. It is not the cold, dead object that evolutionary theorists depict, though it *should* be if it is indeed billions of years old.

Three Stages in Accepting New Scientific Theories

Someone has aptly noted that there are three stages in the acceptance of a new scientific idea by the scientific establishment:

(1) It could not possibly be true.

(2) What difference would it make if it were true?

(3) We knew it was true long ago.

In the case of the transient lunar phenomena, the third stage is hard upon us. Now that a probable source of energy for the transient events is recognized to exist, namely, a hot lunar interior, the events themselves are being legitimized. After more than four centuries of observations, the transient lunar phenomena are finally being accepted into the fabric of conventional science, but not without continuing opposition (Sheeham and Dobbins, 2001).

Footprints

The footprints left by the astronauts will not fade away since there is no wind or liquid water on the moon.

13. When did Crater Bruno form?

This particular lunar crater deserves special comment because of its unusual history. The crater's formation apparently was witnessed on June 18, A.D. 1178, by several Catholic monks. Medieval chronicles relate that "just after sunset, it was reported by five men that the upper horn of a new moon split and from the division point fire, hot coals, and sparks spewed out . . . this phenomenon was repeated a dozen times or more. Then the moon took on a blackish appearance" (DeYoung, 2000). The report apparently describes a major impact on the moon with a resulting explosion, as pictured in Figure 2-4. Great clouds of dust caused a darkening of the moon for days or weeks.

This report of "fire on the moon" is doubted by many astronomers who assume that a large lunar impact occurs only once every million years or so. If collisions are this rare, then the probability of observing such an event within the past 1,000 years is only 0.001. Even the young-looking crater Copernicus is assumed to be at least millions of years old. However, a new crater matching Bruno's description has actually been located on the moon's far side just beyond the Sea of Crises (Mare Crisium, Appendix 1). The Clementine spacecraft in 1999 further showed that this crater, 13 miles in diameter, is very fresh in appearance. Lunar instruments from the Apollo program also record a continuing *ringing* or vibration that may result from the collision event over eight centuries ago (Calame and Mulholland, 1978). The crater has been named for the philosopher Giordano Bruno (1548–1600). Bruno was burned at the stake for heresy by the church hierarchy of his day.

The likely conclusion is that medieval sky watchers did indeed see the impact event resulting in Crater Bruno. This shows that at least some lunar craters are quite recent in age, contradicting the *ancient* formation view of craters. The Clementine spacecraft also photographed debris from recent avalanches and cave-ins along the inside walls of the crater. Such extensive weathering on a short time scale is entirely unexpected. Creationists see the moon as a young, dynamic satellite of the earth. Certainly, most large craters formed in earlier history. However, crater formation may still be 1,000 times greater than usually assumed, one example being Crater Bruno. If true, a creationist prediction is that the moon may be due for another major, observable impact event anytime within the next few centuries.

Figure 2-4. A sketch of what was seen by several witnesses when lunar Crator Giordano Bruno was formed in A.D. 1178.

14. What is the evolutionary future of the moon?

Naturalistic scientists envision a very bleak future for the moon, earth, and the entire universe beyond. The moon will continue to slowly spiral outward from the earth and tides will gradually diminish. Billions of years in the future our moon will be only a dim, distant dot in the night sky. The precious night light and tides will no longer occur on Earth. The moon's revolving time will increase to about 50 days and the earth's rotation time also will slow to this same time period. The revolving moon will become "locked in" above one location on Earth, never to be seen from the earth's opposite side.

As for the sun, it will someday begin to gradually expand and become a *red giant* star, increasing to 50–100 times its present size. In 2–3 billion years the earth's continents will be scorched and the oceans will evaporate away. All life on Earth then will cease. The sun's enormous size will completely vaporize the inner planets including Mercury, Venus, and perhaps Earth. After its red giant phase, the sun will shrink to a small *white dwarf* star, giving off little light. The evolutionary view of the distant future is bleak indeed.

Many astronomers believe the universe itself will continue to expand, with the distant galaxies finally moving out of range of our telescopes. Our own Milky Way galaxy will merge with nearby galaxies such as the Magellan Clouds and Andromeda. Eventually, all lights in the sky will exhaust their fuel and be extinguished. Universe objects gradually will collapse together into black holes. There will be very cold, very dark loneliness everywhere. This finality is called "heat death" with no available energy in the universe. Such a pessimistic view of the future is completely contrary to the creation perspective. Instead, Scripture describes a bright future with an eventual re-creation of a new heaven and earth (Isa. 65:17, 66:22; 2 Pet. 3:13; Rev. 21:1).

MOON MISSIONS
Do additional research on former or future lunar missions. You might consider looking up material online, in books, or watching videos. Write a report about the reasons behind the race for the moon, about the different astronauts and how this affected their families, or about ideas and concepts being addressed concerning a return to the moon.

First Moon Landings

A regular desktop computer of today has five to ten times the computing power of the computers used to land men on the moon. The moon's surface vibrated for 55 minutes with the impact from the *Apollo 12* moon landing.

CHAPTER WORD REVIEW

accretion — a growth of the earth and its moon from dust and gas occurring side-by-side in space

ex nihilo — from no previously existing matter (literally, out of nothing)

planetesimal — a large Mars-sized space object said by some evolutionists to have collided with earth and formed the moon

radioisotope dating method — involves chemical analysis to determine the ratios of parent and daughter atoms in a sample

Roche limit — breakup of a moon occurs within about 2.44 planetary radii of its host planet

yom — Hebrew word translated day

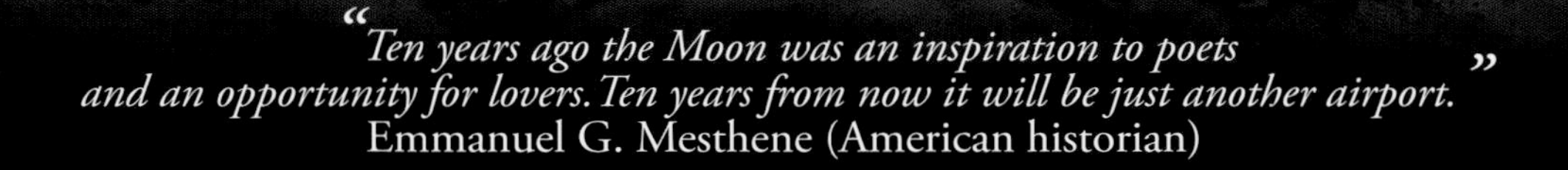

"Ten years ago the Moon was an inspiration to poets and an opportunity for lovers. Ten years from now it will be just another airport."
Emmanuel G. Mesthene (American historian)

3 PURPOSES OF THE MOON

Learning Objectives

1. Describe the value of the moon's reflected light.
2. Assess how the moon has influenced the arts throughout history.
3. Explain the lunar calendar and how the moon determines the dates of Easter.
4. Outline how seasons and weather are affected by the influence of the moon.
5. Analyze how plants and animals are impacted by the moon's influence.
6. Recognize the importance of the ocean tides.
7. Describe how the moon is an energy source for the earth.
8. Discuss the natural resources available on the moon.
9. Assess how the moon protects us from space collisions.
10. Analyze the value of an eclipse.

Words to Recognize

albedo, calendar, corona, harvest moon, vernal equinox

MOON NAMES

Month	English names	Native American names	Other names used
January	Old Moon	Wolf Moon	Moon After Yule, Ice Moon
February	Wolf Moon	Snow Moon	Hunger Moon, Storm Moon, Candles Moon
March	Lenten Moon	Worm Moon	Crow Moon, Crust Moon, Sugar Moon, Sap Moon
April	Egg Moon	Pink Moon	Sprouting Grass Moon, Fish Moon, Seed Moon, Waking Moon
May	Milk Moon	Flower Moon	Corn Planting Moon, Corn Moon, Hare's Moon
June	Flower Moon	Strawberry Moon	Honey Moon, Rose Moon, Hot Moon, Planting Moon
July	Hay Moon	Buck Moon	Thunder Moon, Mead Moon
August	Grain Moon	Sturgeon Moon	Red Moon, Green Corn Moon, Lightning Moon, Dog Moon
September	Corn Moon	Harvest Moon	Corn Moon, Barley Moon
October	Harvest Moon	Hunter's Moon	Travel Moon, Dying Grass Moon, Blood Moon
November	Hunter's Moon	Beaver Moon	Frost Moon, Snow Moon
December	Oak Moon	Cold Moon	Frost Moon, Long Night's Moon, Moon Before Yule

SHOOTING FOR THE MOON
Would You Marry Me?

Honeymoon

The word *honeymoon* comes from the Babylonian tradition of their wedding drink, honey-flavored wine, being provided to the groom by the bride's parents for the month following the wedding.

1. Is the moon a useful night light?

Genesis 1:16 states that "God made two great lights; the greater light to rule the day, and *the lesser light* to rule the night." Our moon actually has no light of its own. It is a reflector of sunlight and indeed dominates the night sky. The amount of moonlight we receive each night varies greatly with its phases. This evening light is pleasant, non-glaring, and sufficient for its purpose.

Throughout history, moonlight has guided evening travelers on land and sea. Also, the moon is especially useful to agriculture. Consider, for example, the "harvest moon." This name is given to the full moon occurring nearest the autumn equinox, around September 21. Due to the tilt of its orbit, the harvest moon rises at nearly the same time on several successive evenings. During the rest of the year, the moon rises about 50 minutes later each night. The harvest moon provides extra light for farmers at the fall harvest time. A compensating delay of the moon's rising occurs in March, six months later. The chart above displays some of the other full moons; their traditional names in American folklore give hints as to the seasonal usefulness of their light.

The full moon surface has a flat appearance. This view is unusual since rounded, illuminated objects commonly have a three-dimensional look. Light reflected from their edge typically is directed away from the observer, giving a darkened border (Figure 3-1). The rough surface of the moon, however, reflects light equally in all directions. This results in the same amount of observed light from nearly every portion of the lunar surface.

The moon has an average reflectivity, or *albedo,* of just 7 percent. This means that about 7 percent of incident sunlight is reflected by the moon. This lunar reflection contrasts with reflection from the earth (31 percent) and Venus (76 percent). If the moon had a larger albedo, its evening glare would be very unpleasant to our eyes. The limited amount of reflected moonlight is sufficient for our needs and provides a soft, attractive light.

Figure 3-1. A comparison of light reflected from smooth and rough rounded surfaces. The moon fits the latter case. A smooth surface results in edge darkening. A rough surface results in uniform reflection.

2. How has the moon influenced the arts?

Over the years, the beautiful moon has inspired much creativity in music, art, and literature. Moonlight sets a mood of silence, graceful movement, and romance. Consider just a few classic references to the moon from English literature:

> The moon, the governess of the floods,
> pale in her anger washes all the air.
> William Shakespeare
> "A Midsummer Night's Dream," 1595

> The moon, rising in clouded majesty . . .
> unveiled her peerless light, and over
> the dark her silver mantle threw.
> John Milton
> "Paradise Lost," 1667

> The moving moon went up the sky,
> And nowhere did abide:
> Softly she was going up,
> And a star or two beside.
> Samuel Taylor Coleridge
> "The Rime of the Ancient Mariner," 1798

> That orbed maiden with white fire laden,
> whom mortals call the moon.
> Percy Bysshe Shelley
> "The Cloud," 1820

The moon has also been used as a theme in musical pieces. These include "Moonlight Sonata," Beethoven's most popular piano sonata (1801), and "Clair de lune" (1905) by Debussy. Several hymns of the Church refer to the moon as a picture of God's faithfulness:

> Thou burning sun with golden beam,
> Thou silver moon with softer gleam.
> O praise him, O praise him!
> Francis of Assisi (1182–1226)
> "All Creatures of Our God and King"

> His kingdom spread from shore to shore,
> Till moons shall wax and wane no more.
> Isaac Watts (1674–1748)
> "Jesus Shall Reign"

> The moon shines full at his command,
> And all the stars obey.
> Isaac Watts (1674–1748)
> "I Sing the Almighty Power of God"

> Summer and winter, and springtime and harvest,
> Sun, moon and stars in their courses above,
> Join with all nature in manifold witness
> To thy great faithfulness, mercy and love.
> Thomas Chisholm (1866–1960)
> "Great Is Thy Faithfulness"

In many ways the moon has enriched our appreciation of *the creation.*

3. What is the lunar calendar?

The word *calendar* comes from Latin, meaning "to call." This historically refers to priests who called people to worship when the new moon (chapter 1) was spotted in the evening sky. Many early cultures, including Old Testament Hebrews, used the lunar calendar. The months began with each new moon phase. The complete set of phases is traversed in 29½ days, so months varied between 29 and 30 days. However, there is not an exact number of lunar cycles during one year on Earth. The lunar year consisted of about 354 days while the year of seasons is 365¼ days long. To reconcile the lunar year, a 13th month was inserted about once every three years. The lunar calendar is still followed by some of the Arab countries, and a crescent moon is commonly pictured on their flags. This "horned moon" has also become a symbol for the religion of Islam (chapter 4).

The lunar calendar is possible because the moon is so predictable in its motion, phases, and eclipses. It exactly obeys the designed laws of orbital motion and celestial mechanics. These relationships give dependability and stability to the universe. Lunar motion has been very useful in the testing of fundamental theories and laws of nature.

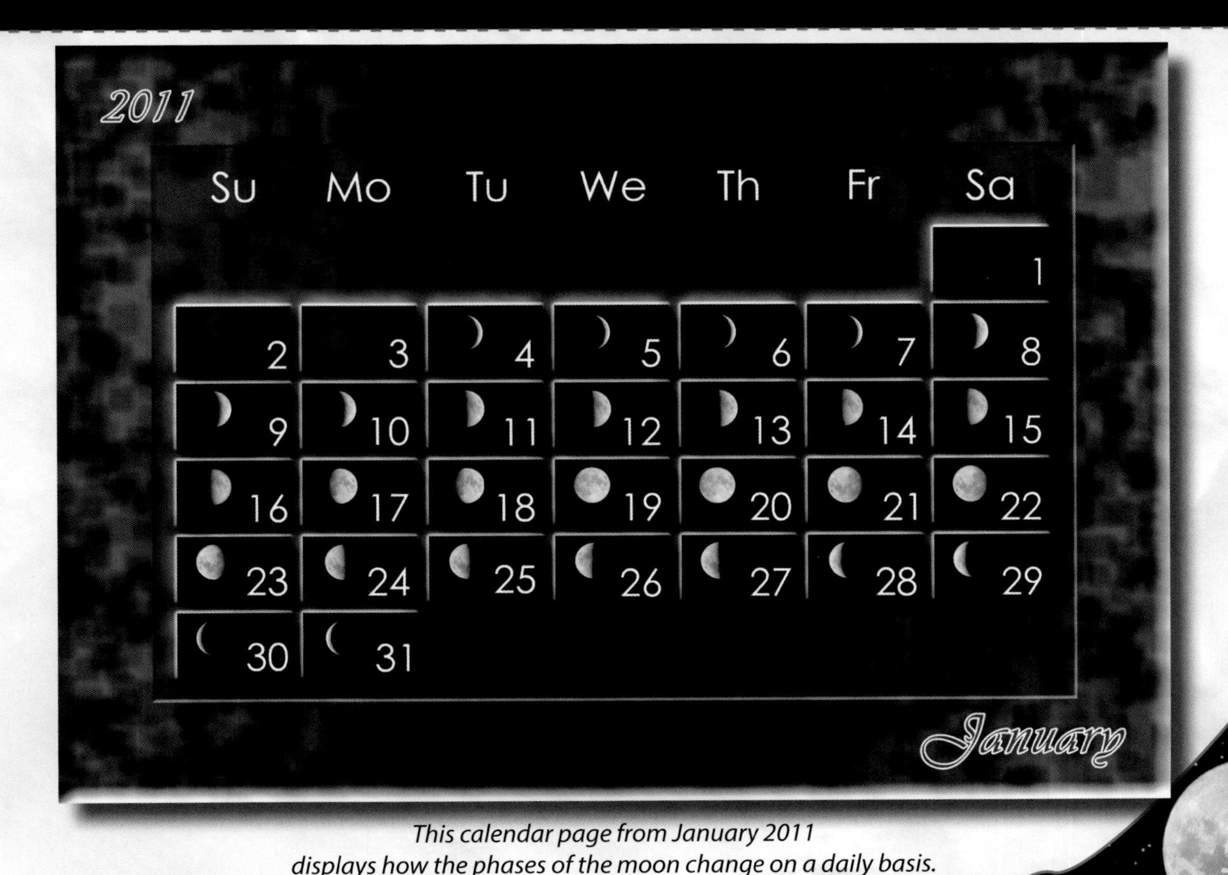

This calendar page from January 2011 displays how the phases of the moon change on a daily basis.

Year	Easter Sunday
2010	April 4
2011	April 24*
2012	April 8
2013	March 31
2014	April 20
2015	April 5
2016	March 27*
2017	April 16
2018	April 1
2019	April 21
2020	April 12
2021	April 4
2022	April 17
2023	April 9
2024	March 31
2025	April 20
2026	April 5
2027	March 28
2028	April 16
2029	April 1

Table 3-1. Easter dates during a twenty-year time span, determined by the full moon phase. Asterisks show the earliest and latest dates in this list.

4. How does the moon determine Easter dates?

Easter commemorates the resurrection of the Lord Jesus. Its date on our calendar was established by Roman Emperor Constantine the Great and the church Council of Nicaea in A.D. 325. To determine Easter for a particular year, look at the calendar and find the date of the *vernal equinox*, the first day of spring, around March 20–22. Now look for the next full moon, often indicated on calendars. Easter will then fall on the following Sunday. According to this formula the earliest possible date for Easter is March 22, next occurring in 2038. The latest date is April 25, in 2285. Most often, Easter comes during the first week of April. Table 3-1 lists Easter dates for several years. It is pleasing to realize that the special Easter celebration of our Lord's victory over death is determined by the created sun and moon. The Easter formula is in agreement with the divinely ordained purpose of heavenly lights as markers for times and seasons (Gen. 1:14).

5. Does the moon affect our seasons and weather?

Earth's seasons are due to the tilt of its rotation axis, relative to the plane of the solar system. This 23½ degree angle tilt of the axis from vertical apparently causes the sun to move back and forth across the equator during the year (Table 3-2). The sun is positioned directly above the equator just two days a year. These are the spring and autumn equinoxes, around March 21 and September 21. Computer studies show that the moon stabilizes the earth's tilt angle (Ward and Brownlee, 2000). The nearby moon acts as a massive counterweight that holds the earth's spin-axis in place. With no moon, the earth's axis would swing erratically between 0 degrees and 90 degrees due to gravity pulls from the sun and other planets, mainly Jupiter. Seasons would then be unpredictable and much more severe. For example, an earth tilt substantially greater than 23½ degrees could lead to a permanent freezing of the oceans and the end of all life. Computer models indicate that this *wandering* of the earth's axis without a moon would be gradual, occurring over thousands of years. Psalm 104:19 correctly states that "He appointed the moon for seasons."

Planet Mars has a tilt and rotation period quite similar to the earth, 24 degrees and 24.5 hours respectively. However, Mars has no large moon. Therefore, the tilt angle of Mars may have changed substantially in historical time. This could possibly explain Mars's erosional evidence from the past, including its apparent dry riverbeds. On Earth, we owe our stable climate largely to the moon.

The phases of the moon were long thought to affect our weather. During the first quarter phase, the visible lunar mares (seas) have names that describe fair weather: Tranquillity *(Tranquillitatis)*, Serenity *(Serenetatis)*, Fertility *(Foecunditatis)*, and Nectar *(Nectaris)*. The third quarter moon was historically associated with stormy weather with these mare names: Storms *(Procellarum)*, Moisture *(Humorum)*,

Clouds *(Nubium)*, and Rains *(Imbium)*. Actual weather records, however, show no such correlation with the first and third quarter moon phases. A full moon *does* provide the earth with slightly additional light and warmth. Careful measurements show that during the full moon, the worldwide average temperature rises by about .01 degree (Balling and Cerveny, 1995).

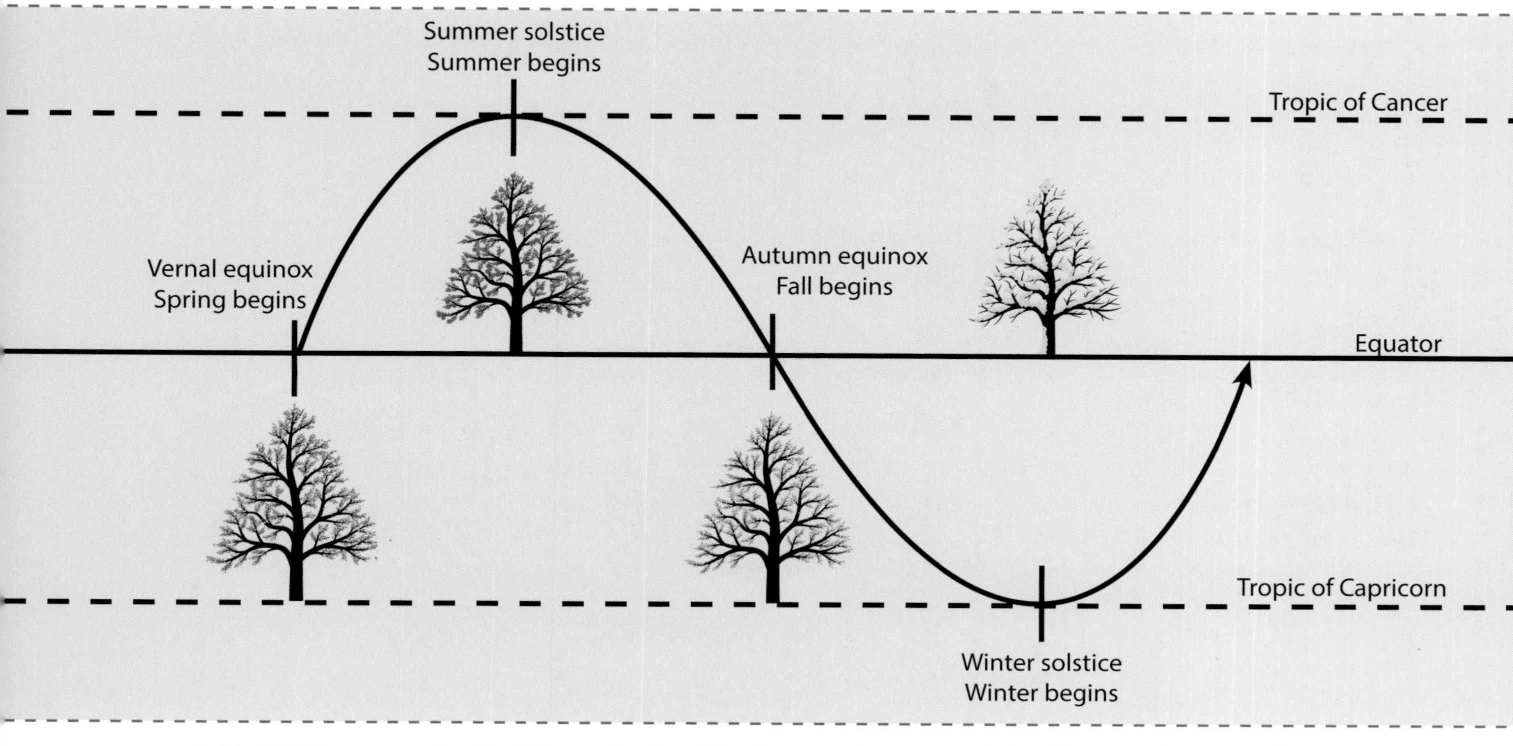

Table 3-2. The apparent path of the sun back and forth across the equator during the year. The Tropic of Cancer is at a latitude of 23.5° above the equator; the tropic of Capricorn is 23.5° below the equator.

6. How does the moon affect plants and animals?

The moon's light and tides have a significant effect on the earth's plants and animals. Farmers are especially familiar with lunar influences. For example, the time of planting and harvesting is often correlated with moon phases. The phases near full moon provide substantially more night light that apparently can help with germination and early plant growth. Animal behavior also appears to be connected with moon phases in unknown ways. This area of practical experience is poorly understood by science.

There are many examples of specific plant and animal behavior connected with the moon. One example is the California *grunion*. During spring and summer these small fish spawn during the full moon phase. At this time of extra high tides, the grunion leave their eggs buried in the sand at the shoreline. Two weeks later, at the next high tide during the new moon phase, the young hatchlings are swept back out to sea. This pattern is also a common reproductive cycle for many other marine animals, including horseshoe crabs. Around the world, the rich biodiversity of intertidal mudflats and mangrove swamps responds to the rhythm of the tides.

Certain fish such as the *moon wrasse* depend on moonlight for illuminating their food at night. Moonlight stimulus is also important in the life cycle of the salmon. These fish undergo cell structure modification during their transition between fresh and salt water. This cell change is found to occur when the moon is new or in its thin crescent phase.

On a shorter time scale, daily tides control the activity of barnacles, crabs, snails, clams, oysters, and shore birds. Successful fishermen know the importance of time of day and the moon phase when fishing in the ocean or inland waters. Fiddler crabs show especially complex behavior with color changes adjusted to sunshine and feeding activity directed by tides.

Many animals apparently use the relative positions of the sun and moon to tell direction. These include fish, turtles, spiders, and insects. Their celestial navigation must take into account the ever-changing orientation of the sun and moon. In addition to the sun and moon, many migrating birds use the stars as a compass. We know very little about the mechanisms, memory, and instinct involved in animal migration.

7. How important are the ocean tides?

Tides and wind together generate the ocean currents worldwide, and studies have shown that tides may dominate the process (Wunsch, 2000). About 3 trillion watts of tidal power are continually dissipated in the oceans. This nonstop power production is roughly equal to that supplied by all the power plants on Earth, including nuclear, fossil fuel, and hydroelectric plants.

Tides are essential to the health of the oceans. The shorelines are continually scrubbed as the water rises and falls in its daily rhythm. In the process, the seawater is oxygenated. Poison runoff from the land to the sea by rivers is diluted, dispersed, and broken down. Without this major stirring of the seas, the water would become stagnant and unhealthy. Sea life would die, especially along the shorelines of the world. This includes the abundant plant life of the oceans. There are vast amounts of "grasses of the sea," or floating plankton, plus larger sea plants such as the "kelp forests" in shallow areas. In fact, there is more total plant life by weight (biomass) in sea water than on all the land. This follows because the world is about 70 percent covered with water. Plants "breathe" the opposite of us, taking in carbon dioxide and giving off oxygen. Therefore, if sea plants perished, the earth's atmosphere would rapidly deteriorate. We then would have insufficient oxygen for life. In this way our very breath is dependent on the lunar tides. Even in a cursed, imperfect world, the ecological design of the world is amazing, including the essential functions of the moon.

8. How is the moon an energy source for the earth?

Research continues on the generation of electricity using lunar tides. The rise and fall of sea water can be used to spin turbines similar to the hydroelectricity produced at power dams on major rivers. One location of successful lunar energy production is on the Rance River in northern France. High tides in the English Channel cause a surge of water moving up the river every 12 hours. As this water returns to the sea at low tide, it is directed through a generating station. This is accomplished with a large-sized, gated dam. The facility produces 240 megawatts of power, about one-quarter that of a large nuclear or fossil fuel power plant. A tidal power station also operates at Annapolis Royal in Nova Scotia, Canada. Tidal electric energy uses no fossil fuels and produces no waste products or heat exhaust. In future years there surely will be an increased interest in tidal generation stations along the coastlines of the world.

9. What natural resources are on the moon?

This question is related to space exploration. Could it be possible that mankind originally had the created potential to visit and manage the moons and planets of the solar system? If true, this option has been largely canceled by limitations brought on by the Curse. Still, it is of interest to consider the vast potential resources in space, including the moon.

As far as we know, the moon has no reserves of gold, silver, or diamonds. The lack of significant lunar water also rules out underground hydrothermal deposits of mineral ores that occur commonly on Earth. Nevertheless, the moon is an object of great interest for exploration. Its two-week length of daylight with no clouds would provide an excellent location for solar energy collectors and telescope observing. Research has shown that lunar soil has adequate minerals for plant growth. On the moon, an enclosure would be necessary to protect plants from the temperature extremes and to provide an artificial atmosphere. The high quartz content of lunar soil possibly could be melted to make glass and ceramic materials. With little water content, the resulting glass would be extremely strong and could easily support building structures. The weak lunar gravity, one-sixth that of earth, would also simplify construction.

The moon's soil and rocks contain the common elements oxygen, silicon, titanium, and iron. Extraction of the oxygen could provide a source of rocket fuel. However, another important fuel component, hydrogen, is rare on the moon.

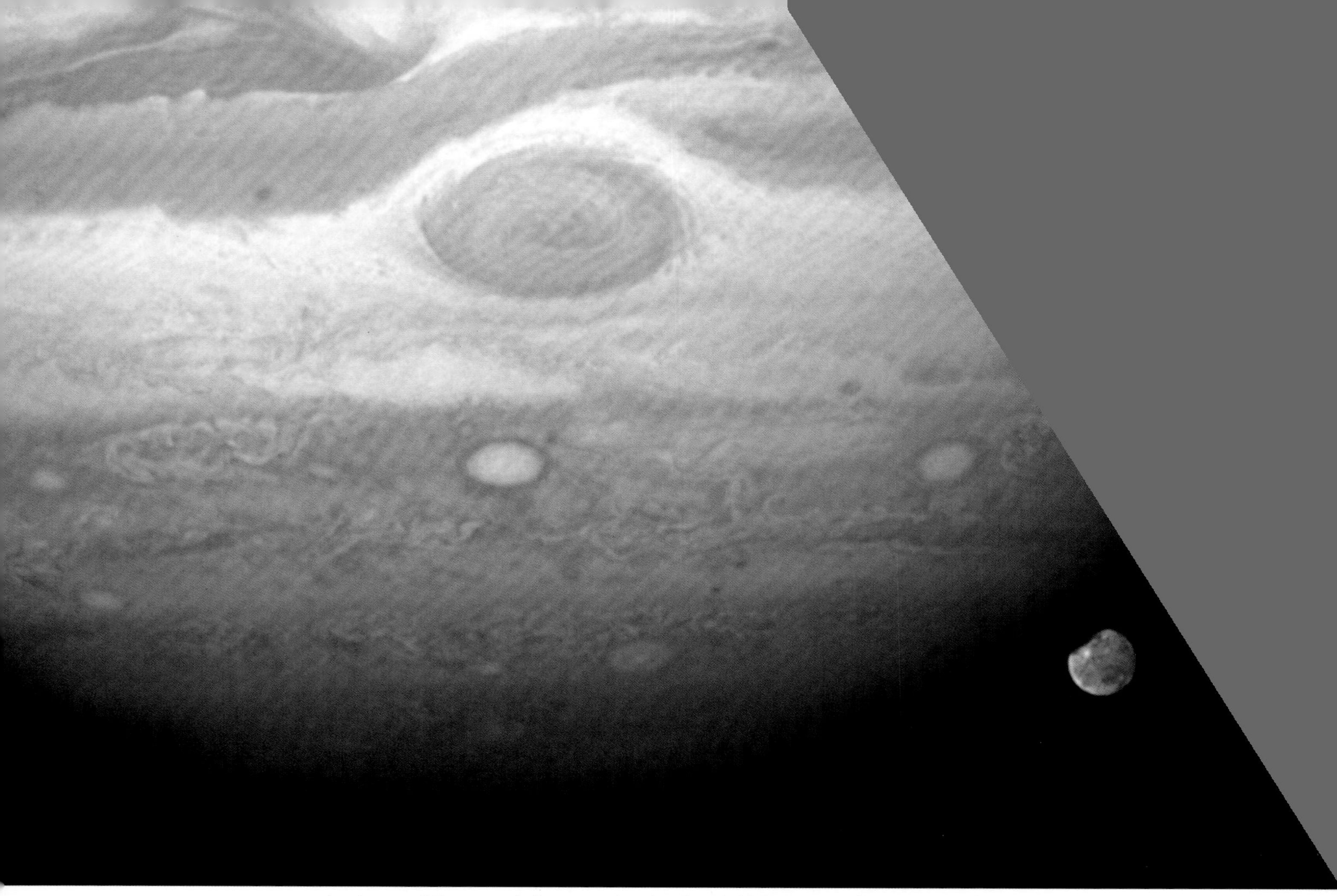

Jupiter and Ganymede image taken by the Hubble Space Telescope (April 9, 2007)

10. How does the moon protect us from space collisions?

Earlier we discussed the doubtful theory of lunar origin by an ancient planetary collision with the earth (chapter 2). In contrast, however, the created moon actually *prevents* space collisions with the earth (Comins, 1991). Many space rocks that would otherwise hit the earth are instead drawn to the moon by its gravity attraction. The far side of the moon, exposed to incoming objects, is especially heavily cratered. Some of these lunar craters are over 150 miles (241 km) in diameter. Such large impacts on Earth could cause profound changes in the earth's atmosphere and climate.

On a larger scale, the planet Jupiter also protects the earth from collisions with asteroids and comets. Jupiter's mass is greater than the other eight planets combined. As a result, Jupiter has a large attractive gravity force for objects approaching the inner solar system. In 1994 astronomers watched fragments from Comet Shoemaker-Levy strike Jupiter. The approaching comet broke up when it crossed within Jupiter's Roche limit (chapter 2). Two dozen separate impact collisions then occurred on Jupiter, each one much more energetic than a nuclear blast. Had any of the comet fragments hit the earth, a crater at least 50 miles in diameter would have resulted.

Blue Moons

This originally referred to an extra full moon that would occur in a season (based on solstices and equinoxes). There are approximately 12.37 full moons a year, so a blue moon comes around about every 2.7 years. Some refer to it as a month that has two full moons.

11. What is the value of an eclipse?

Eclipses declare God's glory by their rarity, precise prediction, and their beauty. A total solar eclipse is one of the most amazing sights in nature. As the moment of totality approaches, twilight comes and confused birds give evening calls. The deep shadow of the moon hurtles across the land at over 1,000 miles an hour. Darkness then arrives like a light that has been switched off. Stars appear overhead and the outer atmosphere of the sun, the *corona*, glows around the edge of the moon's disk. With the sun blocked, air temperature quickly drops by several degrees. Minutes later, the sun begins to uncover and daylight appears once again.

Solar eclipses have proven valuable in many ways. An eclipse provides information on the sun's *corona*, its outer atmosphere of hot gases. Precise eclipse measurements reveal the sun's exact diameter. Eclipses also help date the past. Many ancient historical records include mention of eclipses. This data unlocks the exact chronology and thus the true history of a large segment of the first millennium B.C. The significance of the eclipse data for biblical studies is incomparably great, for it provides confirmation, unavailable for well over two thousand years, that the chronological systems employed by Old Testament scribes were perfectly accurate (Thiele, 1965; Whitcomb, 1977).

One eclipse description is found in Amos 8:9: "And it shall come to pass in that day," says the Lord God, "that I will make the sun go down at noon, and I will darken the earth in broad daylight." The date for Amos is thought to coincide with two eclipses that occurred in 784 and 763 B.C. Amos's hearers could thus relate to the prophecy of coming darkness. Other probable eclipse references include Micah 3:6 (716 B.C.) and Jeremiah 15:9 (610 B.C.).

THE MOON IN ART

Artists often depict the moon in paintings and will sometimes make it larger to exaggerate its size and place in their art. Research paintings and seek out different ways artists represent their works.

Moon Rotation

The moon makes one rotation around Earth every 27 days, 7 hours, and 43 minutes, and moves around Earth in an oval shape. (The moon is actually egg-shaped, with the large end pointed to Earth). The earth's equator rotates at about 1,000 miles per hour while the moon equator rotates at just 10 mph.

CHAPTER WORD REVIEW

albedo — name given to the average reflectivity of the moon

calendar — Latin word meaning "to call"

corona — the outer atmosphere of the sun made of hot gases

harvest moon — name given to the full moon occurring nearest the autumn equinox, around September 21

vernal equinox — the first day of spring, around March 20–22

"As we flew into space, we had a new sense of ourselves, of the Earth, and of the nearness of God."
James Irwin (Apollo 15 lunar module pilot)

4 THE MOON IN SCRIPTURE

Learning Objectives

1. Define what the moon symbolizes in the Bible.
2. Discuss the double-revelation theory.
3. Explain what language of appearance is in Scripture.
4. Contrast the biblical account of the moon's creation with ancient pagan ideas.
5. Relate what the Old Testament says about the worship of the moon.
6. Assess the worship of the moon in the ancient Near East and compare to sun and moon worship today.
7. Articulate how the Curse and the Flood may have affected the moon.
8. Analyze whether the sun and moon really stopped during Joshua's long day.
9. Illustrate the destiny of the moon.

Words to Recognize

cosmogony, cosmology, Enuma Elish, ex nihilo

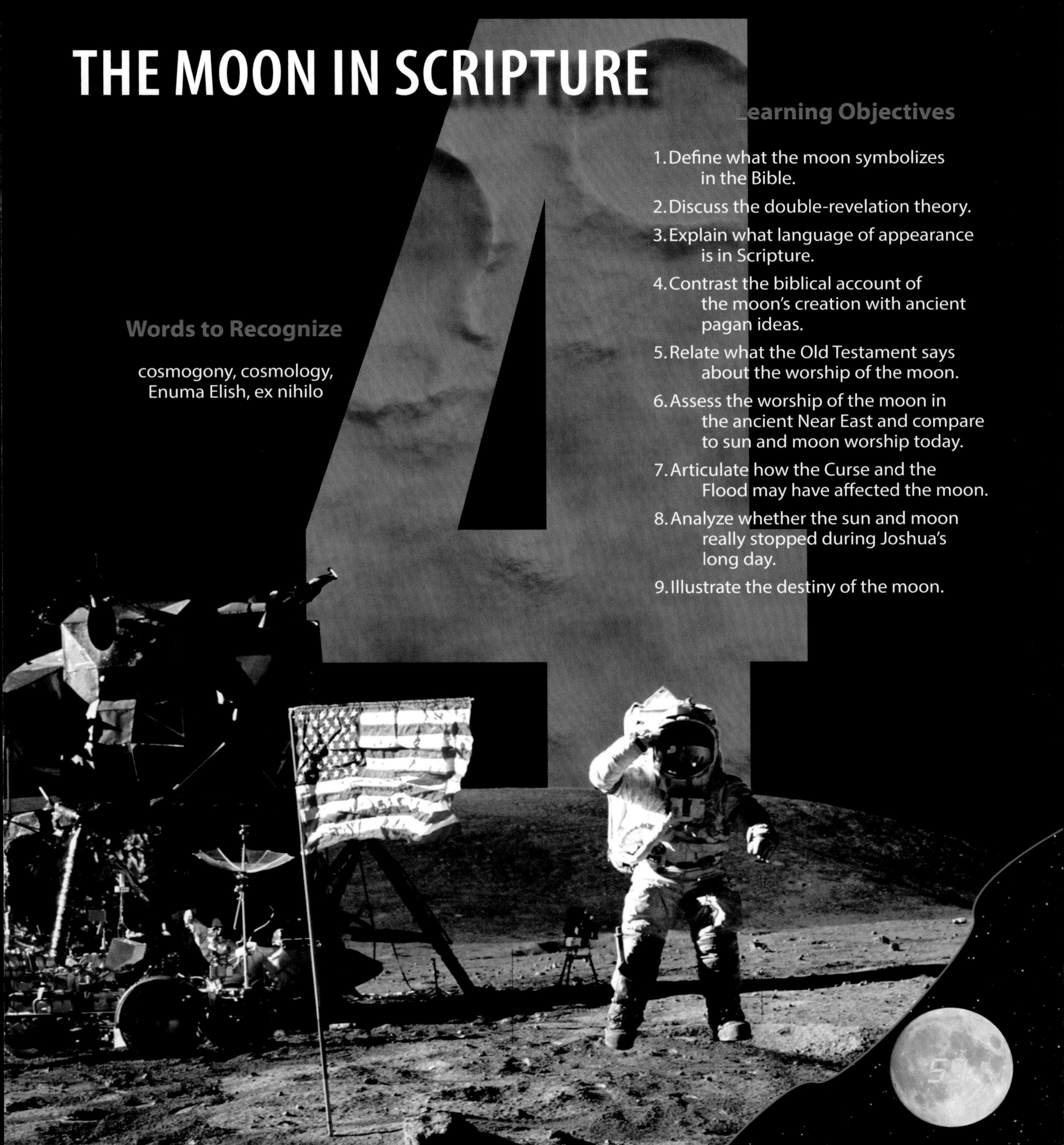

1. What does the moon symbolize biblically?

The sun, moon, and stars show God's *faithfulness*. Jeremiah 31:35–36 declares, "Thus says the Lord, who gives the sun for a light by day, the ordinances of the moon and the stars for a light by night. . . . The Lord of hosts is His name: "If those ordinances depart from before Me, says the Lord, then the seed of Israel shall also cease from being a nation before Me forever." The permanence of the moon is also declared in Psalm 72:5, 89:37, and 121:6.

Solomon, probably the wisest naturalist between Adam and Christ (1 Kings 4:29–34), referred to the sun and moon as models of performance and dependability when he spoke of worship in the future Kingdom Age: "They shall fear You as long as the sun and moon endure, throughout all generations" (Ps. 72:5). Likewise, Ethan the Ezrahite, a contemporary of Solomon (1 Kings 4:31), wrote that the Davidic throne would be "established forever like the moon, even like the faithful witness in the sky" (Ps. 89:37).

The heavenly lights are also a symbol of God's great *glory*. Biblical writers freely expressed their fascination with the spectacular beauty of the moon. Four thousand years ago, Job referred to "the moon moving in brightness" across the night sky (Job 31:26). A thousand years later, Solomon described his lovely Shulamite bride-to-be as "fair as the moon" (Song of Sol. 6:10). A thousand years later, the Apostle Paul observed that "there is one glory of the sun, another glory of the moon" (1 Cor. 15:41), thus agreeing with the Psalmist that "the heavens declare the glory of God" (Ps. 19:1). Future changes in the moon as a symbol of God's *judgment* are described in chapter 4.

2. What is the double-revelation theory?

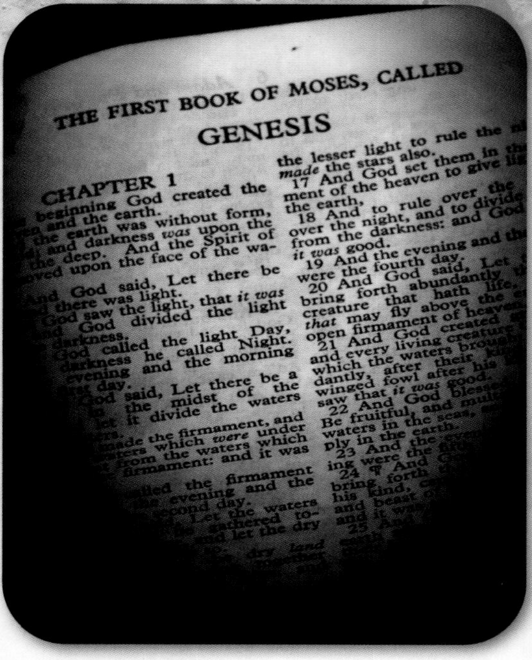

This theory maintains that God has given man two distinct and ultimate revelations of truth, each of which is fully authoritative in its own realm: the revelation of God in *Scripture* (i.e., special or biblical revelation) and the revelation of God in *nature* (i.e., general or natural revelation). Although these two revelations differ greatly in their character and scope, they cannot appear to intelligent men to contradict each other since they are given by the same self-consistent God of truth. The theologian is the God-appointed interpreter of Scripture, and the scientist is the God-appointed interpreter of nature, each having specialized lenses for reading the true message of the particular "book of revelation" that he has been called upon to study.

Those who endorse the double-revelation theory also maintain that whenever there is an apparent conflict between the conclusions of the scientist and the theologian, especially with regard to such questions as the origin of the universe, the solar system, the moon, the earth, plant and animal life, or man, the theologian must rethink his interpretation of Scripture at these points. Scripture must be brought into harmony with the consensus of scientists, since the Bible is not a textbook of science and these problems overlap the territory in which science alone must give us detailed and authoritative answers.

It is held that this approach is necessary, because if a grammatical/historical interpretation of the biblical account of the creation of the moon, for example, should lead the Bible student to adopt conclusions that are contrary to the prevailing views of trained scientists concerning the origin and nature of the material universe, then he would be guilty of making God a deceiver of mankind in these vitally important matters. But a God of truth cannot lie. Therefore, the Genesis creation account must be interpreted in such a way as to bring it into agreement with the generally accepted views of contemporary scientists. The early chapters of Genesis, we are told, were written primarily to give us answers only to such "spiritual" questions as "Who?" and "Why?" Scientists, however, must answer the important questions, "When?" and "How?" Similar words from four centuries ago are attributed to Johann Kepler: "The Bible tells us how to go to heaven, not how the heavens go." Similar sentiments have been expressed by astronomer Hugh Ross: "The facts of nature may be likened to a sixty-seventh book of the Bible" (Ross, 1994). The implication is that current science understanding is an updated supplement to biblical truth.

Though the double-revelation theory has gained considerable popularity in some Christian circles, it fails to come to grips with major theological and scientific realities. The following discussion will focus on this theory:

1. **Proponents of the theory do not recognize the tremendous limitations that inhibit the scientific method when applied to the study of ultimate origins.**

In the very nature of the case, the scientific method (which analyzes the processes of nature in observable and repeatable events) is incapable of coping with once-for-all and utterly unique events, or even the moral and spiritual realities that give significance to human endeavor. It ultimately fails when an attempt is made to employ it in analyzing the supernatural and miraculous acts of God, which form the foundation pieces of the Judeo-Christian worldview.

Those who exclusively employ the scientific method in historical sciences ignore the possible anti-theistic bias of the scientist himself as he handles the facts of nature in arriving at a *cosmology* (i.e., a theory concerning the basic structure and character of the universe) and a *cosmogony* (i.e., a theory concerning the origin of the universe and its parts). Ultimately, they fail to point to the only true source for understanding the mysteries of the universe.

2. **Advocates of the double-revelation theory overlook the scientific problems that plague currently popular naturalistic/evolutionary theories concerning the origin of the material universe and of living things.**

The impossibility of explaining mechanistically the formation of the biosphere that surrounds us at such close range on planet Earth continues to be an embarrassment to evolutionists.

The thermodynamic and mathematical barriers to a chance transition from non-life to life in a primeval sea, the debilitating and even lethal effects of the vast majority of mutations, the large and as yet unbridged gaps between the various kinds of plants and animals found in the fossil record, and the clear evidence of global catastrophes (rather than generally uniform processes) in the formation of coal seams and other fossil strata have all contributed to the comparatively recent, widespread reappraisal of Lyellian/neo-Darwinian models of geology and paleontology.

3. **Most proponents of the double-revelation theory underestimate and even deny the supreme authority and self-evident clarity of God's special revelation in Scripture.**

The biblical record of physical and biological origins is thoroughly historical and amazingly detailed. Genesis 1–11 is accurate and authoritative history. Therefore, pre-Abrahamic biblical history stands firmly upon the rock of objectivity and cannot be twisted to suit the whim of the interpreter. By making so many detailed statements about history, chronology, geography, astronomy, geology, and zoology, the Author of Genesis is showing His hand as one who expects to be taken seriously and who actually provides material suitable for investigation. This is in sharp contrast to the sacred writings of other ancient religions.

Biblically informed Christians, therefore, must look with profound dismay at the double-revelation theory and other attempted "harmonizations" of Genesis and evolutionism in its various forms. Scientists who desire to find answers to the questions of ultimate origins, meaning, and destiny cannot succeed by positioning themselves in opposition to God's infallible Word. Nor can theologians! Therefore, scientists and theologians who seek for truth in these vitally important areas of research must function within the universally valid guidelines of spiritual and natural laws. Scripture and science are *not* mutually contradictory, and they are *not* in competition with each other. They have been designed by God in such a way that the latter is defined, conditioned, and guided by the former, resulting in genuine scientific truth. The harmony is rather between Genesis and objective, empirical science. It is in this realm of reality that the supposed warfare between science and theology comes to an end.

Consequently, Christians must abandon all hope of formulating a scientifically valid cosmogony if they fall prey to the popular notion that science provides an *independent* and *equally authoritative* source of information with the Bible concerning the creation of the universe, the solar system, and the earth with its living forms, and that science alone is competent to tell us *when* and *how* such things occurred, while the Bible merely informs us in "spiritual" terms (or, as neo-orthodox theologians would express it, "suprahistorical" language) as to *who* brought the universe into existence and *why.*

The truth of the matter is that the Word of God not only provides us with the only reliable source of information as to the *who* and the *why* of these great events, but also provides essential information concerning the *when* and *how*. Such an assertion has, of course, been widely resisted in modern times, usually with ominous references to Luther's rejection of Copernicus and the pope's persecution of Galileo. But in these famous (and thankfully rare) examples of opposition to genuine scientific discovery in the name of Scripture, a more careful examination of the facts reveals that in these and similar cases the Bible's own guidelines to its interpretation had been neglected.

3. What is *language of appearance* in Scripture?

From Genesis to Revelation, the Bible consistently and thankfully avoids (for the sake of effective communication to mankind) highly technical teaching of scientific data or concepts. Nevertheless, the Bible provides perfectly accurate descriptions of things by the use of the language of appearance. A perfect example of this principle is the account of the creation of the sun and moon in Genesis 1:16. "Then God made two great lights: the greater light to rule the day, and the lesser light to rule the night. He made the stars also." On an absolute scale, of course, the sun and moon are not "great lights" compared to many of the giant stars. In fact, the moon is not a "light" at all in the sense that the sun is a light. But from the perspective of earth-dwellers, the statement is vastly more meaningful than a technical astronomical analysis. Furthermore, the statement is perfectly accurate. There are only two great lights visible to the unaided eye, not three or ten.

Another outstanding example of this principle is found in Revelation 7:1, which speaks of the earth having "four corners." This does not suggest that the Bible subscribes to the flat-earth concept, for the following phrase explains that "the four corners" refer to "the four winds of the earth." Even today, meteorologists use the four directions of the compass to describe wind movements, without thereby implying that the earth is flat.

Finally, and perhaps best known of all, are biblical statements that refer to "the rising of the sun" (e.g., Rev. 7:2; NASB; Ps. 19:4–6). Does this mean that the Scriptures teach geocentrism? Not at all, for this is a language of appearance so appropriate that it cannot be improved upon even by astronomers or the common language of our day.

4. How does the biblical account of the moon's creation compare with ancient pagan ideas?

The Genesis creation account stands in stark contrast to other ancient origin stories. To illustrate this fact, consider the *Enuma Elish* ("When on high"), a creation myth written on seven clay tablets in Mesopotamia about 1800 B.C. This was soon after Abraham was called by God to leave the same area (Gen. 12; chapter 4). This epic is a significant expression of pagan religion in the Akkadian language. The story begins with two primordial gods, the husband *Apsu* (sweet water) and wife *Tiamat* (salt water). They had many children or younger gods who were very noisome. Because of this, Apsu decided to slay them. Instead, however, one of his sons killed him. Then another son called *Marduk* killed his mother, Tiamat. She was divided into two parts "like a shellfish," which became the heavens above, including the moon, and the earth below. Marduk also made people, as slaves, to do the work here on Earth (Pritchard, 1969). In contrast to the *Enuma Elish*, how majestic and dignified is the Genesis account of origins! Biblical creation is uniquely different from the *Enuma Elish* and all other secular creation stories.

5. What does the Old Testament say about moon worship?

Assuming a date for the Book of Job around 2100 B.C., we have in this remarkable document the earliest inspired record of man's thinking about the moon. Bildad the Shuhite asked Job, "If even the moon has no brightness and the stars are not pure in His sight, how much less man. . . ." (Job 25:5–6; NASB; also see Job 15:15). Job himself confirmed this concept of the infinite superiority of God to the moon, and insisted that worship of the sun or moon would rightly incur judgment because of the blasphemy involved in such idolatry.

If I have observed the sun when it shines, or the moon moving in brightness, so that my heart has been secretly enticed, and my mouth has kissed my hand [in the homage of worship]: this also would be an iniquity deserving of judgment, for I should have denied God who is above (Job 31:26–28).

Several centuries later (c. 1405 B.C.), when Moses was preparing Israel for entrance into the land of Canaan, stern warnings were issued concerning the desperate danger of being influenced by the solar, lunar, and astral worship cults of the nations that had lived there for centuries: "Lest you lift up your eyes to heaven, and when you see the sun, the moon, and the stars, all the host of heaven, you feel driven to worship them, and serve them, which the LORD your God has given to all the peoples under the whole heaven as a heritage" (Deut. 4:19; see also Deut. 17:3 and Amos 5:26).

After the establishment of the theocratic kingdom under David, God made a promise to His people that has caused some perplexity: "Behold, He who keeps Israel shall neither slumber nor sleep. The LORD is your keeper; the LORD is your shade at your right hand. The sun shall not strike you by day, *nor the moon by night*. The LORD shall preserve you from *all evil*" (Ps. 121:4–7, emphasis added). While it is true that sunstroke was at times a special danger in the Near East (2 Kings 4:18–20; Jon. 4:8; Isa. 49:10), the heat of the sun could hardly qualify as one of Israel's major threats, to say nothing of "the moon by night." It seems more appropriate, therefore, to consider this promise within the context of sun and moon worship. Idolatrous people were presumably afraid that a neglect of proper sacrifice to these heavenly deities would result in their being somehow smitten during the day by the sun or during the night by the moon.

About 700 B.C., seven centuries after Israel had entered the Promised Land, crescent ornaments (Isa. 3:18), shaped like the moon and directly inviting worship of the moon (see Judges 8:21, 26), were popular with "the daughters of Zion" whose lifestyles were influenced by Near Eastern idolatry more than by their Lord.

Soon after the days of King Hezekiah and Isaiah the prophet, during the seventh century B.C., celestial idolatry swept into Judah like a flood. King Manasseh (690–640 B.C.) actually "worshipped all the host of heaven and served them . . . and he built altars for all the host of heaven in the two courts of the house of the LORD" (2 Kings 21:3–5). When Josiah inherited the throne, he attempted to purge the land of these influences and especially "the idolatrous priests . . . who burned incense to Baal, to the sun, to the moon, to the constellations, and to all the host of heaven" (2 Kings 23:5).

But Josiah's reforms were too superficial and too late. Radical forms of idolatry had deeply infected the hearts of the vast majority of Jews. Therefore, Josiah's great contemporary, the "weeping prophet" Jeremiah (see Jer. 9:1, 13:17), was commanded by God to cease praying for the nation (Jer. 7:16; 11:14; 14:11). Judah was almost totally corrupted by submitting to the all-encompassing influence of Near Eastern idolatry, including the worship of the moon, and would therefore be destroyed and deported by the cruel armies of Nebuchadnezzar, the king of Babylon. (See Jer. 8:1–2; 19:13).

What the warnings of Moses, Isaiah, and Jeremiah did not accomplish for the people of Israel, the Babylonian captivity *did* accomplish, at least superficially. The Jews became known henceforth as "the people of the Book," and gross idolatry such as moon worship receded into a solemn memory of their pre-exilic past, enshrined forever in the inspired records of the Old Testament.

Moon Gravity

The moon's gravity is one-sixth that of Earth, meaning that you weigh considerably less on the moon's surface.

6. How prevalent was moon worship in the Ancient Near East?

What exactly were the enormous influences toward the worship of the moon that so threatened Job and his contemporaries in the late third millennium B.C. and the people of Israel for over 800 years after their exodus from Egypt?

The science of archaeology has demonstrated the deification of the moon from early Sumerian times (third millennium B.C.) to Islamic times throughout western Asia, the cradle of postdiluvian civilization. The great Sumerian city of Ur, in lower Mesopotamia, was especially devoted to the worship of the moon, under the name of *Nanna* or *Nannar*, long before the time of Abraham. Surprisingly, *Nanna* was thought to be the father of the sun-god *Shamash,* though he himself was the son of a yet greater god, *Enlil.*

Soon after 2000 B.C., the moon temple at Ur was greatly enlarged by Ur-Nammu, first king of the great Third Dynasty of Ur. Fourteen hundred years later, King Nabonidus of Babylon restored this temple. A cuneiform inscription has been found there in which Nabonidus concluded with a dedication to Nannar, lord of the gods of heaven and earth, and a prayer for the life of himself and of his son Belshazzar. This temple, or ziggurat, remains today as the best preserved of the many that can still be seen in Mesopotamia.

The Akkadians called the moon-god by the name *Sin*, "the lamp of heaven and earth," "the king of all gods," and "the Divine Crescent." The crescent is the familiar symbol for *Sin* in Mesopotamian art. When the far-off Hittites appealed to the Sun-god in their hours of special need, they referred to him as "the favorite son of Sin." The great Hammurabi (c. 1700 B.C.) introduced himself in his famous Code as "the descendant of royalty, whom Sin begat."

It is fascinating to trace the presence of moon worship along the routes followed by Abraham and his descendants. Called by the true Creator of the moon to leave his homeland, Abraham came to the city of Harran in northern Mesopotamia, today on the southern border of southeastern Turkey. To judge from the later prominence of moon worship in this city, the "Divine Crescent" was probably honored in his day, too. Like the Apostle Paul who received a similar call from the Creator of heaven and earth over 2,000 years later, Abraham's spirit must have been "provoked within him as he was beholding the city full of idols" (Acts 17:16; NASB, in reference to Athens).

Leaving Harran about 2090 B.C., Abraham continued southward toward Canaan, doubtless encountering signs of the Ugaritic moon deity along the way, whose name was *Yarikh*. At the city of Hazor, north of the Sea of Galilee, archaeologists have discovered the remains of an ancient Canaanite shrine dedicated to the full and crescent moons (see page 67). Proceeding on into Egypt because of a famine in Canaan (to be followed later by Jacob's entire family), he may well have learned of *Khonsu*, the moon-god of Thebes.

As millions of Abraham's descendants re-entered the Promised Land under the leadership of Joshua, they were confronted with the fortress-city of Jericho, which, as its very name indicates (*yᵉrîhô*, compare *yārēāh*, "moon"), was probably dedicated to the Semitic moon-god.

Eight centuries later, as the Jews were deported to Babylon because of their persistent indulgence in moon worship (among other reasons), it is fascinating to ponder that moon worship played a significant role also in the final collapse of the neo-Babylonian empire. Because his mother, Adad-guppi, was a devoted priestess of the moon-god *Sin* at Harran, King Nabonidus (who married a daughter of Nebuchadnezzar) clashed constantly with the Marduk priests in his capital city of Babylon because of his insistence on introducing moon worship there. Finally, in 553 B.C., he left Babylon, "entrusted the kingship" to his son Belshazzar, and moved to Tema in northwestern Arabia because it was an ancient center of worship of the moon-god.

This royal and priestly clash over the issue of moon worship resulted in Babylon being left in the hands of Belshazzar, an utter profligate (see Dan. 5), and thus open to the assault of Cyrus the Great and his Medo-Persian armies in 539 B.C.

The prominence of moon worship in the experience of Abraham and his descendants down to the time of Cyrus (c. 539 B.C.) serves to clarify the message of the first chapter of Genesis that was given to Israel through Moses just before they returned to Canaan. The message was that the moon (as well as all other astronomical bodies) was *not* a god to be worshiped. In fact, so far from being a deity, it was not even the offspring of a god or goddess (as was claimed for *Nanna/Sin*). Instead — and this is utterly and permanently devastating to all forms of moon worship — it was created instantly, *ex nihilo,* and subsequent to an earth already carpeted with vegetation by the incomparable God of Israel (Gen. 1:11–19).

In conclusion, it must be recognized that if the modern theistic-evolution interpretation of the origin of the moon had been adopted by ancient Israel, the message of Genesis would have been reduced to the level of just one more Near Eastern mythological cosmogony with little or no impact upon idolatry-prone ancients. This profoundly important fact must be weighed carefully by theologians who are tempted to mold the opening chapter of the Bible into conformity with currently popular naturalistic theories of origins.

7. How prevalent is sun and moon worship today?

There has been a modern revival of occultic practices involving the sun, moon, and stars. This includes traditional astrology and various *new-age* beliefs. There is a long list of Internet sites and book titles that are dedicated to the pagan worship of sky objects. These sources offer a mixture of pseudoscience and cultic nonsense. Thus far, fortunately, these fringe groups have had little impact on society.

In astrology, the moon is considered an important factor in determining a person's personality. Special significance results if the moon is located in one's constellation or horoscope sign at the time of birth. Healthwise, the moon has traditionally been said to influence one's nerves, nutrition, emotions, mental state, and fertility. However, all of these astrology ideas are entirely groundless and non-scientific. In fact, Scripture strongly warns against delving into astrology. Modern witchcraft, likewise condemned in Scripture, worships the moon as a symbol of power and fertility.

Let us consider this question of nature worship more broadly and apply it to the sun. The origin of life is commonly attributed to solar energy, which somehow sparked life into primordial Earth chemicals. All life today is dependent on warmth, energy, and food provided by the sun. Secular astronomers also predict that the sun will someday expand to a *red giant* phase and snuff out life on Earth (chapter 2). The sun alone is therefore looked at as the originator, sustainer, and ultimate extinguisher of life on Earth. In this way the sun has become, by default, the god of mankind. This book has been written, in harmony with God's written revelation in Scripture, to put the moon and sun in their proper place. That is, they are merely secondary objects created by God for our benefit, not for our control.

Moon Worship

A foot-high pillar engraved with hands raised toward the full and crescent moon. Discovered in 1956 in excavations at Hazor, northern Israel, and dating to about 1400 B.C.

8. Did the Curse or the Flood affect the moon?

The rebellion of mankind is described in Genesis 3:1–13. The resulting judgment included a curse upon the serpent, pain for mothers in childbirth, thorns and thistles, painful toil, and, most significantly, spiritual and physical death. From the biblical perspective, catastrophes such as lunar impact craters *may* have occurred at the Fall or at other times in history.

Some creationists suggest that major cratering of the earth and moon are related to the universal deluge that almost destroyed humanity and involved the collapse of the antediluvian vapor canopy and the upheaval of the "fountains of the great deep" (Gen. 7:11). This may have involved an original tenth planet between Mars and Jupiter where the asteroid belt is now located. The planet may have somehow exploded, showering the solar system with fragments. This then resulted in many lunar craters. On the earth, smaller particles became precipitation nuclei, which aided the formation of raindrops.

Several large craters also have been found buried *within* flood deposits on Earth. Other surface craters, such as the Barringer Crater of northern Arizona, formed at some time after the Flood. This later time frame also may be true of the fresh-rayed craters observed on the moon. It must be stated, however, that no objective evidence for the absolute age of cratering has yet been discovered either in Scripture or in astronomical or geological science.

Moon Madness

People often correlate full moons with fanatical behavior or madness, thus the words *lunatic* and *lunacy* were created with this in mind.

9. Did the sun and moon really stop moving during Joshua's long day?

Joshua 10 describes a battle between Joshua's forces and the Amorites of Canaan. After marching from Gilgal all night, Joshua and his men were challenged by the great South Palestine confederation of Canaanite armies. During the struggle Joshua spoke in the name of the Lord saying:

"Sun, stand still over Gibeon;
And moon, in the Valley of Aijalon"
(Josh. 10:12).

Joshua's prayer then was answered,

> *So the sun stood still,*
> *And the moon stopped,*
> *Till the people had revenge*
> *Upon their enemies* (Josh. 10:13).

The additional daylight, perhaps 12 hours extra, allowed Joshua to defeat the enemy. The moon was included because it was also visible at the time. This was surely one of the most fascinating miracles of the Old Testament (Davis and Whitcomb, 1989).

The event did indeed occur exactly as recorded. Joshua 10:14 declares that the day was unique in its time extension. The event also was commemorated in the *Book of Jasher* (Josh. 10:13), a volume lost to history and no longer available. There are no other known annals available that describe this time-delay miracle. Various explanations have been offered to explain the long day, including a solar eclipse, the use of nonliteral poetic language, or perhaps an alteration of the earth's rotation. Scientific objections to such a miracle, namely, that the stopping of Earth's rotation would cause mountains, seas, and people to fly off into space, are entirely irrelevant. God certainly has power to control all the ancillary materials and forces involved when performing miracles. As always, it simply is futile to attempt a scientific analysis of miracles. They are, after all, *super*natural. The point is that God *did* answer Joshua's prayer. The sun, moon, and stars are subservient to the Creator who placed them in the sky in the first place.

The Bible, however, has provided us with certain important guidelines for understanding "Joshua's long day." When the Flood ended, God promised to Noah that "while the earth remains . . . day and night shall not cease" (Gen. 8:22). This was confirmed 800 years after the time of Joshua: "If you can break My covenant with the day and My covenant with the night, so that there will not be day and night in their season, then My covenant may also be broken with David My servant" (Jer. 33:20–21; see also 33:25–26 and 31:36). In other words, God has promised that the earth would not cease rotating on its axis at the present rate until the very end of human history. Thus, it is biblically illegitimate to "multiply miracles" when the Bible teaches "an economy of miracles."

What, then, did happen during Joshua's long day? Since his need was a prolongation of light — not a slowing down of the earth's rotation — his need could have been met by a supernatural continuation of sunlight and moonlight in the central portion of the Promised Land for "about a whole day" until his army could follow up its great victory and completely destroy the enemy.

A similar, and equally spectacular, miracle of light refraction occurred 700 years later in the days of King Hezekiah (2 Kings 20:1–11). In the courtyard of his palace there was apparently a series of steps (not necessarily a sundial as we would think of it) so arranged that the shadow cast by the sun would give an approximation of the time of day. At the request of this godly king, and doubtless in the presence of a large group of officials (and perhaps foreign ambassadors), the shadow moved backward ten steps (or "degrees").

How did God choose to accomplish this great miracle? Did He cause the earth to stop its rotation and turn backward a little? All true believers would agree that God could have done such a thing, for by Him all things consist or hold together (Col. 1:17). But the Bible makes it rather clear that this was not His method; for in referring to this miracle, 2 Chronicles 32:24 states that Hezekiah "prayed to the Lord; and He spoke unto him and gave him a sign [Hebrew: *mopheth*]." But in verse 31 we are told that the Babylonians sent ambassadors to Hezekiah "to inquire about the wonder [*mopheth*] that was done in the land." Obviously, then, it was a geographically localized miracle that did not involve a reversal of the earth's rotation, with shadows retreating ten degrees all over the Near East. Instead, the miracle occurred only "in the land" (of Judea); and, to be even more specific, it was only in the king's courtyard that "the sun returned ten degrees on the dial by which it had gone down" (Isa. 38:8).

There is a popular story that the long day of Joshua, and also the time extension in Hezekiah's day, have been verified by modern computer studies. It is said that the sun and moon are found to be somewhat out of their proper positions. Adjustments from the time miracles of Joshua and Hezekiah then remove the discrepancy. However, this story is not true. Computers are unable to verify such historical events because the exact prior positions of the sun and moon are not known. Without this information it is impossible to measure an alteration of sun and moon positions. There *are* limits to what computers can do! The Old Testament miracles are true and do not need technical verification.

Moon Craters and More

Over 500,000 craters on the moon are visible from Earth. The deepest craters are over 15,000 feet. The highest mountains are over 16,000 feet. The moon's largest crater, called "Bailly" or "fields of ruin," is about 26,000 square miles.

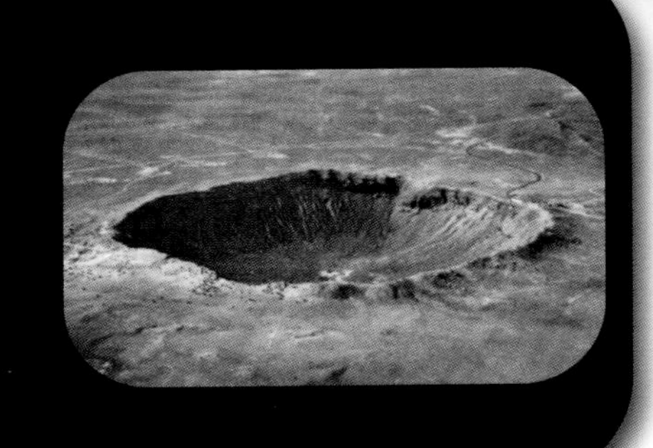

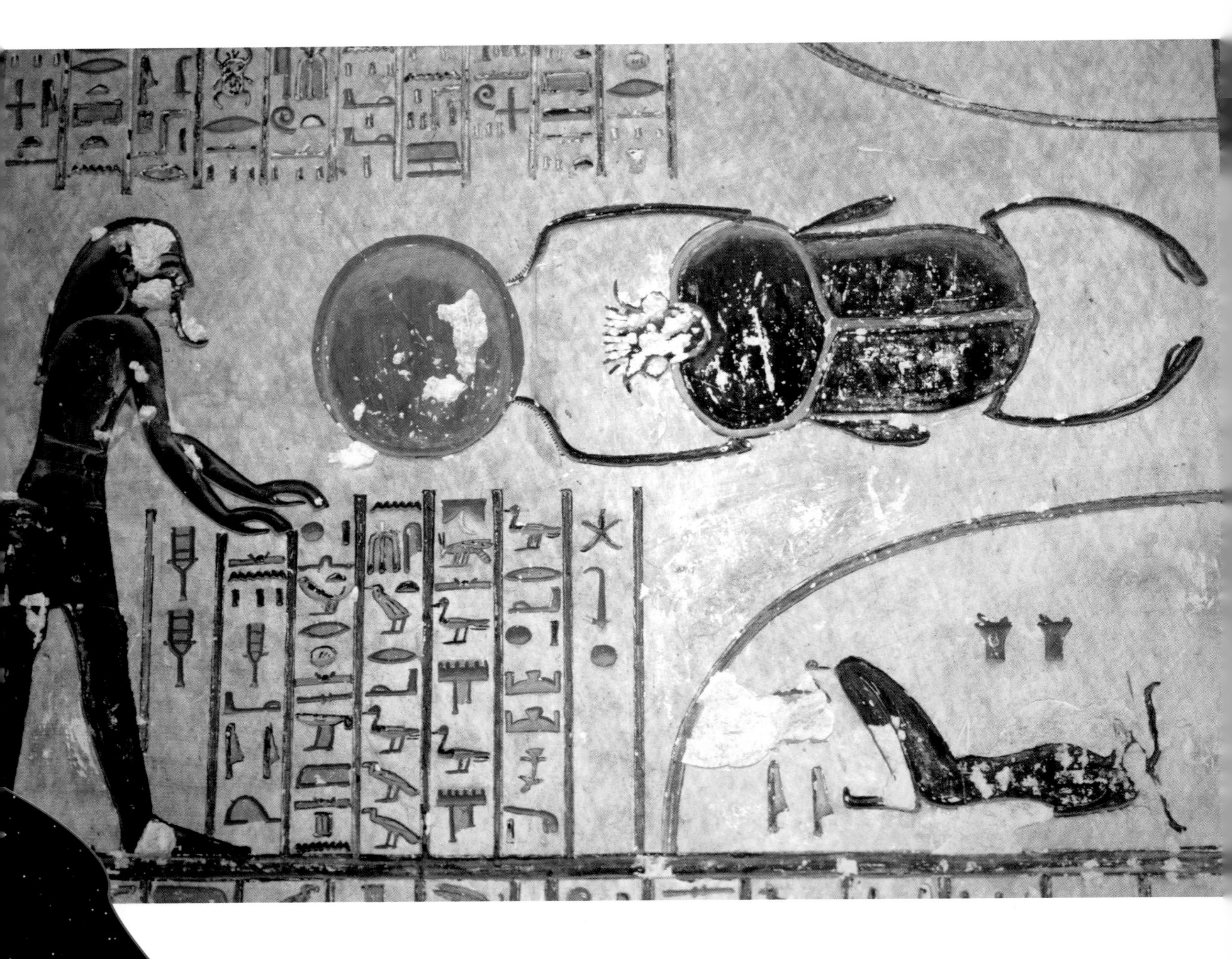

Figure 4-3. One Egyptian concept of stellar movements was that a divine scarab beetle pushed the sun in its assumed orbit around the earth.

10. What is the destiny of the moon?

Since only God can behold future events with perfect accuracy, it is fascinating to search His Word concerning the destiny of the moon. Three future stages may be seen: (1) the obscuring of its light during the Great Tribulation; (2) its availability for Israel's calendar of worship during the thousand-year Kingdom of the Lord Jesus Christ; and (3) its non-existence during the eternal state following the destruction of the heavens and the earth.

First, He told us that "there will be great tribulation, such as has not been since the beginning of the world until this time, no, nor ever shall be" (Matt. 24:21), and that "the moon will not give its light" (Matt. 24:29; as predicted by Isaiah 13:10 and Ezekiel 32:7). From the perspective of earth dwellers, in the midst of the dust and smoke of collapsing cities (see Isaiah 2:10–21), the moon will look "like blood" (Rev. 6:12), and its brightness will be reduced by a third (Rev. 8:12). This was predicted by the prophets Joel (2:10, 31) and Isaiah (24:23) and quoted by the Apostle Peter (Acts 2:20) .

Second, the moon will serve as an essential clock/calendar for Israel during the thousand-year kingdom of Christ on Earth following His second coming. During the present Church Age, observing "a festival or new moon" is not required (Col. 2:16); but in the Kingdom Age, monthly festivals (Ezek. 45:18, 21, 25) and new moon observances (Ezek. 45:17; 46:1, 3, 6) will be prominent aspects of world worship. Isaiah predicted that "from one New Moon to another . . . all flesh shall come to worship before Me" (Isa. 66:23).

Third, the moon will cease to exist at the end of the Kingdom Age. The Apostle John saw in a vision "a great white throne and Him who sat on it, from whose face the earth and the heaven fled away. And there was found no place for them" (Rev. 20:11). Then, John tells us, "I saw a new heaven and a new earth, for the first heaven and the first earth had passed away" (Rev. 21:1). As for the New Jerusalem of the eternal state, "the city had no need of the sun or of the moon to shine in it, for the glory of God illuminated it" (Rev. 21:23). This would be the fulfillment of Isaiah's prophecy, "The sun shall no longer be your light by day, nor for brightness shall the moon give light to you; but the LORD will be to you an everlasting light, and your God your glory" (Isa. 60:19).

Thus, the moon will have served its divine purpose for all mankind when it is darkened during the Tribulation and recognized during the Millennium. During that Kingdom Age, for which our Lord taught us to pray ("Thy kingdom come . . ." Matt. 6:10), Solomon foresaw that "in His days the righteous shall flourish, and abundance of peace, until the moon is no more" (Ps. 72:7).

THE MOON IN SCRIPTURE
Using a Bible concordance, look up the word *moon* in Scripture. Develop a chart regarding how many books mention the word, what specific passages mention it, in what context the word is used, and the total number of usages in the Old and New Testaments.

Moon Names

In our solar system, all the moons are derived from Greek and Roman myths, with the exception of Uranus, the moons of which are derived from Shakespearean characters.

CHAPTER WORD REVIEW

cosmogony — a theory concerning the origin of the universe and its parts

cosmology — a theory concerning the basic structure and character of the universe

Enuma Elish — Mesopotamian creation myth written on seven clay tablets

ex nihilo — a Latin term for creation meaning "out of nothing." God called material things into existence by His Word.

"For when I look at the Moon I do not see a hostile, empty world. I see the radiant body where man has taken his first steps into a frontier that will never end."
David R. Scott (Commander *Apollo 15*)

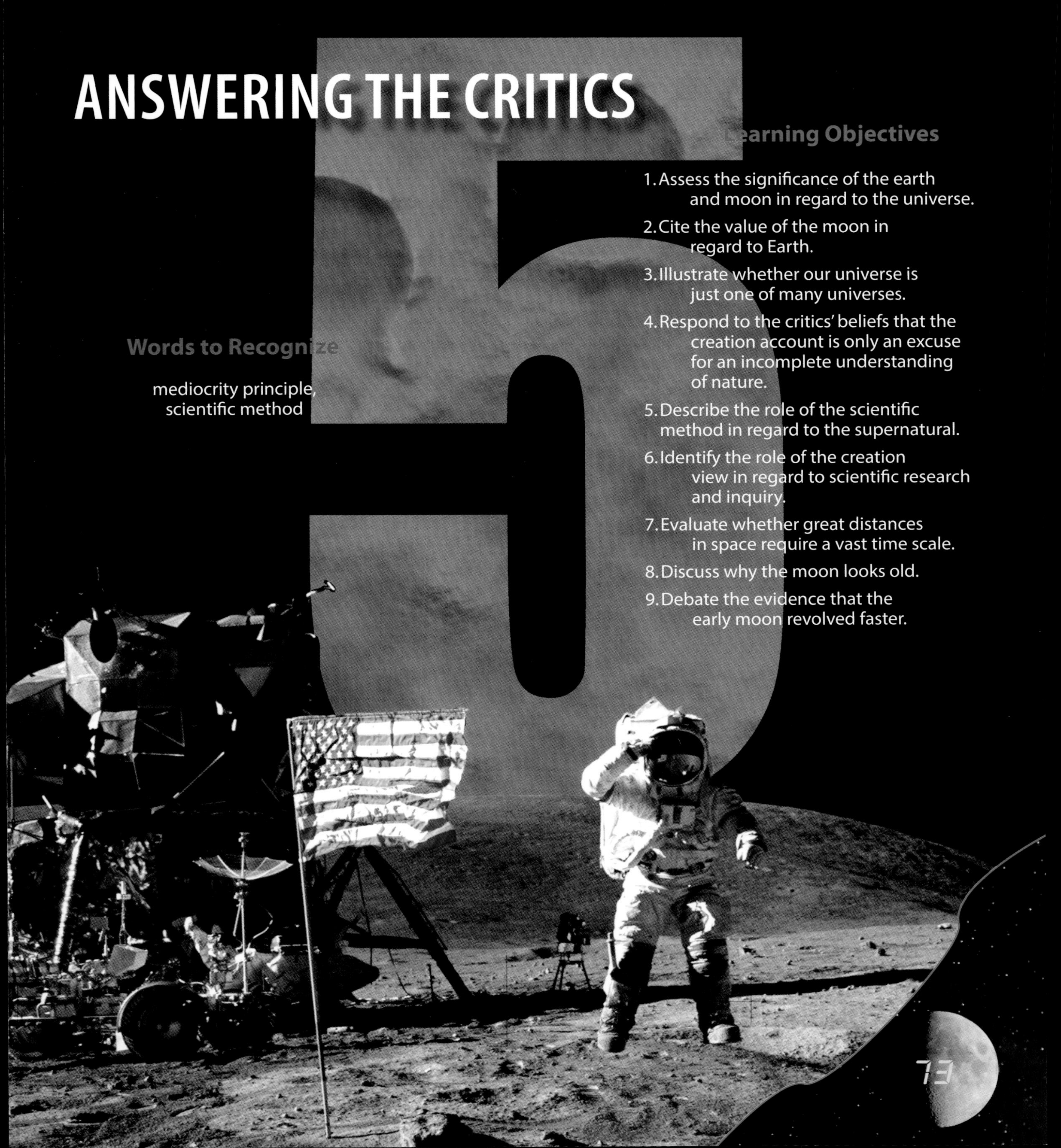

5 ANSWERING THE CRITICS

Learning Objectives

1. Assess the significance of the earth and moon in regard to the universe.
2. Cite the value of the moon in regard to Earth.
3. Illustrate whether our universe is just one of many universes.
4. Respond to the critics' beliefs that the creation account is only an excuse for an incomplete understanding of nature.
5. Describe the role of the scientific method in regard to the supernatural.
6. Identify the role of the creation view in regard to scientific research and inquiry.
7. Evaluate whether great distances in space require a vast time scale.
8. Discuss why the moon looks old.
9. Debate the evidence that the early moon revolved faster.

Words to Recognize

mediocrity principle, scientific method

1. Are the earth and moon insignificant specks in the universe?

It is popular in astronomy literature to deny any special significance for the earth or moon. This emphasis on terrestrial irrelevance follows from the assumption of a chance origin of the earth, moon, and life itself. It is stressed that the earth is "just" an average-sized planet within the solar system. The solar system, in turn, is located in the distant outer region of the spiral Milky Way Galaxy. If the galaxy were reduced to the size of North America, then our entire solar system would be no larger than a teacup. Often connected with this view is the *mediocrity principle*. This is the assumption that there is nothing unique about the evolution of life on Earth, and life is probably duplicated in countless forms in distant regions of space.

In contrast, Scripture gives great physical and spiritual significance to the earth. Planet Earth was created three days *before* the sun, moon, and stars. The divine purpose of the stars relates directly to the earth: to provide signs, a calendar system, illumination (Gen. 1:14), and to declare God's glory to mankind (Ps. 19:1). Planet Earth is also a universal reference point in that Christ walked here among men, and will one day return again. Also, an unseen spiritual battle between the forces of good and evil focuses on this earth and extends to high places (Eph. 6:12).

Some creationists suggest that the earth is positioned at the physical center of the universe. This may be true, but presently there is no way of verification. The exact location of the earth and moon in space is less important than God's special attention, which has been directed toward the earth throughout its history.

2. Would we be better off without the moon?

In 1991 a tabloid headline appeared saying, "Scientists Plan to Blow Up the Moon." This bizarre idea originated with American mathematics professor Alexander Abian, who taught that the moon had a deleterious effect on the earth's weather. He claimed that destruction of the moon would somehow eliminate severe climates and world hunger. Similar ideas have been expressed by a group of Russian scientists. However, the book you are now holding demonstrates the exact opposite: the moon is essential to our health and welfare (see chapter 3). Thankfully, the tabloid headline is an impossible idea. Current technology could not destroy the moon even if it was attempted. Scientists have considered the consequences for the earth if the moon did not exist (Comins, 1993). They conclude that, on a long time scale, the earth would have a rapid rotation, high winds, severe seasons, and a poisonous atmosphere. In other words, without the moon, we could not exist. So much for the misguided suggestion to eliminate the moon!

3. Is our universe just one of multiple universes?

This question comes in response to the abundant design evidence in nature. It is suggested that our known universe is just one of an infinite number of other universes. Therefore, the providential design of the moon, earth, and physical constants is meaningless. After all, there must be many other universes where the conditions are less hospitable to life. With a sufficient number of universes, at least one of them is bound to be ideal for the evolution and development of life, and we happen to live in it.

In response, it should be realized that the multiple universe idea is entirely unprovable and unscientific. Our single universe is all that we can possibly know about. The suggestion of additional universes beyond ours is a desperate attempt to increase the sample size and to manufacture a large statistical group in which our universe is just one member. Theory and mathematics may explore the multi-universe concept; however, this is not reality but instead is *metaphysics* or philosophy. Concerning known universes, the sample size is *one*. The physical universe we inhabit is all we know of, is all that Scripture describes, and it clearly reveals creative design.

4. Is creation an excuse for our incomplete understanding of nature?

Critics often complain that the creation story is a *cover-up* for our lack of science understanding. Since we do not yet understand mechanisms such as the origin of life, we simply say that God did it. This is sometimes called "God-of-the-gaps" reasoning. Critics further predict that the appeal to the Creator eventually will be eliminated, as gaps in our understanding are gradually filled in with further science progress.

Two comments are in order.

First, it is certainly true that science has "closed the gap" in many areas of knowledge. For example, it was once thought that comets in the night sky were supernatural omens of evil. It is now known that comets are natural objects with predictable solar system orbits. At the same time, however, every scientific discovery opens up new fundamental gaps in our understanding. Regarding comets, for example, how did the dependable laws of orbital motion arise? How did comets form, and are there really vast numbers of comets in the outer realms of the solar system? In other words, over time the number of gaps in our knowledge increases rather than decreases.

Second, evolution theory includes some very fundamental gaps that remain forever beyond natural explanation. One example is the mystery of the spontaneous origin of life. All attempts to replicate life or its DNA component using raw materials have failed. Atheistic theories also fail to explain human consciousness, love, and the universal desire of people to worship a higher power (Eccles. 3:11). Such "gaps" in scientific understanding are to be expected by creationists. After all, a supernatural origin lies beyond science inquiry. Consider these supporting references:

> *The secret things belong to the* Lord *our God, but those things which are revealed belong to us and to our children forever, that we may do all the words of this law* (Deut. 29:29).

He does great things past finding out, yes, wonders without number (Job 9:10).

Where were you when I laid the foundations of the earth? Tell me, if you have understanding (Job 38:4).

It is the glory of God to conceal a matter, but the glory of kings to search out a matter (Prov. 25:2).

"For my thoughts are not your thoughts, nor are your ways my ways," says the LORD. "For as the heavens are higher than the earth, so are my ways higher than your ways, and my thoughts than your thoughts" (Isa. 55:8–9).

Oh, the depth of the riches both of the wisdom and knowledge of God! How unsearchable are his judgments, and his ways past finding out! For who has known the mind of the Lord? Or who has become his counselor? (Rom. 11:33–34).

For now we see in a mirror, dimly, but then face to face. Now I know in part, but then I shall know just as I also am known (1 Cor. 13:12).

When it comes to the origin of life, the intricate design in nature, or creation from nothing, it is entirely adequate to say "God did it" and no further explanation is needed (Snoke, 2001).

5. Does the scientific method rule out any appeal to the supernatural?

The scientific method is a plan of action often used in problem solving. The method can be summarized in four steps.

1. Understand the problem.
2. Predict a solution.
3. Carry out this solution.
4. Is the problem solved? If not, return to step 2.

Notice that there is nothing magical about the scientific method. It is simply an approach to problem solving that we all use in everyday experiences.

Science study today is extremely naturalistic. The *supernatural* has been redefined as *superstition*. However, science is not *required* to be antagonistic toward the supernatural in this way. The historical meaning of science (from the Latin *scientia*) is simply *knowledge*. Many of the best-known astronomers have held great respect for creation and spiritual truth. Some of the names are listed in Table 5-1. Their testimonies show that the scientific method is *not* at all in conflict with the supernatural. Instead, the spiritual dimension lies entirely beyond the scientific method. Scientific analysis alone cannot give a complete picture of origins, history, or the future.

6. Is the creation view opposed to science research and inquiry?

One critic has stated that the ultimate goal of creationists is to close down all the research laboratories and simply say, "God did it." But nothing could be further from the truth. Instead, we have a biblical mandate to study the creation so we can better understand its details. The Genesis 1:28 command is to "subdue the earth." This certainly includes the study of nature's details so that we can better manage the earth, bring out its potential blessings, and care for it. Acknowledgment of the Creator of the universe is the best possible starting point for scientific investigation and progress. Consider these references:

And to man he said,
"Behold, the fear of the Lord,
that is wisdom,
And to depart from evil is understanding."

Job 28:28

The fear of the LORD is the
beginning of wisdom;
A good understanding have
all those who do His commandments.
His praise endures forever.

Psalm 111:10

The fear of the LORD is the
beginning of knowledge,
but fools despise wisdom and instruction.

Proverbs 1:7

The fear of the LORD is the
beginning of wisdom,
and knowledge of the Holy One
is understanding.

Proverbs 9:10

The fear of the LORD is the
instruction of wisdom,
and before honor is humility.

Proverbs 15:33

Let us hear the conclusion
of the whole matter:
Fear God and keep his commandments,
for this is man's all.

Ecclesiastes 12:13

Science research proceeds today at a rapid pace. Each year there are about one million new technical articles published. Most of these are read by only a few highly specialized experts. This flood of data and analysis is almost always given a completely secular interpretation. And yet, all research into nature is in truth *creation research*, whether this fact is recognized or not. Everyone has the same data, but the interpretations vary greatly. The purpose of this book is to show the excitement of applying science data to just one area of study, our created moon.

7. Do great distances in space require a vast time scale?

Distances in space extend outward for billions of light years, each light year being about 6 trillion miles (9.5 trillion km) in length. How then can we possibly see far distant objects if the universe is young and recently created? It would seem that there hasn't yet been time for the distant light to reach us.

In the big-bang theory, distance and time are directly connected. The expansion of the universe to its present vast size from a concentrated initial point requires billions of years. However, the creation view does not require this constraint of a gradual big-bang expansion. Instead, the vast universe was instantly and supernaturally formed. In this way the light from distant galaxies reached the earth immediately. This concept of a mature, fully functioning universe is consistent with creation. Notice that God said, "Let there be lights in the firmament of the heavens . . . and it was so" (Gen. 1:14–15).

There are several other possible explanations for seeing distant stars in a youthful universe, aside from a mature creation. One suggestion is that light traveled much faster in the early universe. In this way distant starlight arrived here in a short time. Measurements of light do not clearly show a changing speed today. However, the physics community has found possible indications of a slightly more rapid light speed in the distant past (Weiss, 2001). Perhaps the speed of light was infinite at the moment of creation, when stars were made, and then directly adjusted to its present, lower, constant value.

Another approach to explaining the visible universe involves separate clocks for the earth and outer space. Relativity theory and experimental data indicate that time itself is a quantity that can be *contracted* or *stretched*. This variation in the passage of time becomes significant at high speeds or in the vicinity of large mass. The possibility thus exists that, while normal 24-hour creation days took place on the earth, vast ages transpired in space, and light traveled great distances. This idea of *relativistic time dilation* has been popularized by creationist Russell Humphreys (1994) and also by astronomer Gerald Schroeder. Caution is needed here regarding the validity of inserting present-day physics theory into the supernatural creation week. At some point, the *natural* and *supernatural* must be mutually exclusive.

There is much talk today about additional spatial dimensions beyond the familiar dimensions of length, width, height, and time. Some physics theories predict seven or more additional, unseen dimensions invisibly "folded" within space. This conclusion follows from elegant mathematical equations and models that attempt to describe the nature of the universe. In unknown ways, higher dimensions could be consistent with a rapid formation of the universe and an initial near-infinite light speed.

In the end, it must be realized that God's ways are "past finding out" (Job 9:10; see also Job 5:9). Regarding the particular question under discussion, seeing the distant stars, the challenge for the creationist is to carefully distinguish between the concepts of distance and time. They are entirely separate quantities, and great distance does *not* require a vast timescale.

Name	Specialty	Comments
Fabricus, David 1564–1617	Variable stars	Was a Dutch Reformed pastor in addition to being an astronomer.
Galileo Galilei 1564–1642	Physics	Believed biblical truth.
Keckerman, Bartholomew 1571–1609	Comets	Saw comets as a sign from God.
Kepler, Johann 1571–1630	Planetary motion	Concerning his discoveries, he wrote, "God has passed before me in the grandeur of His ways."
Wendelin, Gottfried 1580–1667	Planets	An ordained priest, he held a deep faith in the Creator.
Gassendi, Pierre 1592–1655	Planetary motion	Taught that God created atoms in a single stroke.
Newton, Isaac 1642–1727	Physics	Wrote, "Our system of planets ... could only proceed from . . . an intelligent and powerful being."
Bradley, James 1693–1762	Stellar motion	Held a strong Christian faith.
Ferguson, James 1710–1776	Instruments	Gave God credit for design in nature.
Wright, Thomas 1711–1786	Milky Way	Taught that religion alone could explain the cosmos.
Herschel, William 1738–1822	Double stars	Wrote, "The undevout astronomer must be mad."
Herschel, Caroline 1750–1848	Comets	Wrote her own epitaph "[She] followed to a better life her father, Isaac Herschel."
Herschel, John 1792–1871	Nebulae	A devout Christian.
Mitchell, Maria 1818–1889	Comets	Wrote, "Every formula is a hymn of praise to God."
Riemann, Bernhard 1826–1866	Mathematics	Defended the Book of Genesis.
Maunder, Edward 1851–1928	Sun	Defended the Bible's accuracy in science matters.
Leavitt, Henrietta 1868–1921	Variable stars	Known for her sincere Christian life and character.
Eddington, Arthur 1882–1944	Stellar interiors	Wrote that the spiritual realm was as real as nature.
Braun, Wernher von 1912–1977	Rockets	Wrote that space "confirms our belief in the certainty of its Creator."

Table 5-1. A partial list of pioneer astronomers who supported creation, listed in order of birth year.

8. If the moon is young, why does it look old?

This question is frequently asked about the earth's age. How can the recent-creation position possibly be defended in the light of our surroundings? We see deep valleys, eroded mountain ranges, and thick sediment deposits. Many of our national parks with their cliffs and valleys are described as "monuments to time." On the moon, as well, we see craters from the distant past, soil that has been pulverized by many impacts, and rounded hills that are regarded as "ancient."

We respectfully suggest that these many features are not ancient, nor do they actually look old when inspected carefully. The important factor in aging is the *rate of change* in the past. Although the earth and moon may show only small changes today, this does not mean present rates have always existed. With the earth, events such as the Genesis flood rapidly altered the entire surface of the planet. Instead of an ancient appearance, the earth's crust can be interpreted as greatly disturbed by the global Flood. This includes tectonic activity, worldwide flood deposits, and the fossil record.

Aside from craters, lava flows, and dust accumulation, the moon's surface may appear much as it did at the time of creation. That is, it may well have been formed with its rolling hills, highlands, and low areas. To say that the moon *looks* old is to show a prior assumption of a long time scale. As a comparison, the Garden of Eden on Earth was surely created with soil and full-grown trees. There was an immediate appearance of age or maturity.

A further answer to this question of appearance concerns rapid changes occurring on the earth. For example, the Mount St. Helens volcano of Washington state erupted in 1980. Observers were astounded at the near-instantaneous change of the landscape for many miles around. This included massive mudflows, explosive destruction of forests, and rapid erosion by moving water. Then, in the following years, a rapid healing of the land began. Vegetation grew back and animal life returned. A visitor today might guess that the volcanic eruption took place centuries ago, followed by a gradual recovery. Such events as Mount St. Helens show how misleading landscape appearance can seem, whether on the earth or moon.

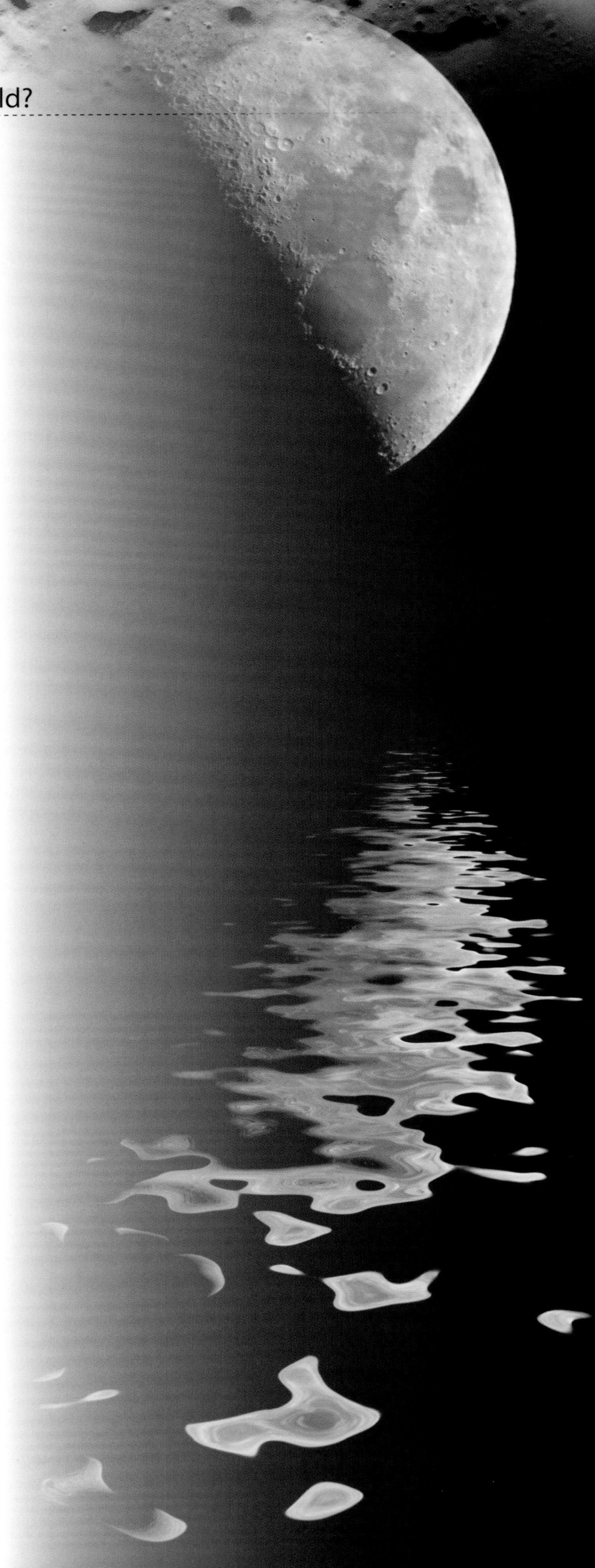

Moon Beliefs

A popular, non-serious idea from the 16th and 17th centuries was that the moon was made of green cheese.

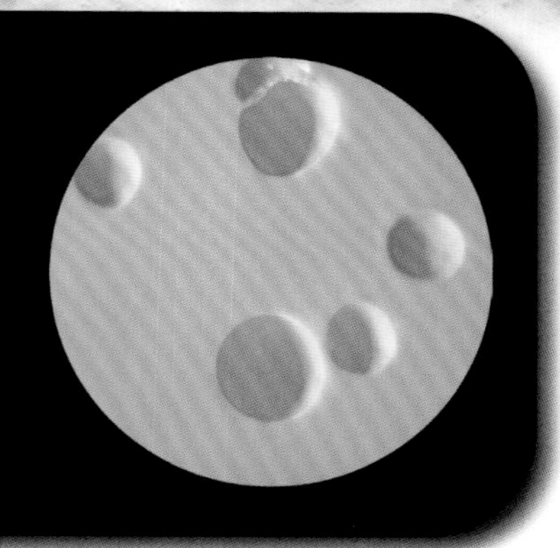

9. How can the majority of scientists be wrong about the moon's origin?

It may sound arrogant to suggest that most scientists are radically wrong in their views of origins, earth history, moon age, etc. After all, how can thousands of professionals worldwide possibly be mistaken? We hold firmly to the recent-creation view, but with humility rather than arrogance. We believe that more emphasis needs to be given to the *tentativeness* of science pronouncements. In fact, considering the history of science, the majority of scientists have *always* been wrong. The philosopher Thomas Kuhn wrote about this phenomenon in 1962. He describes how a particular science theory or belief system grows in popularity until it becomes the "standard view." But its lifetime is limited. Sooner or later a new theory arises and replaces the original. Sometimes the change is minor, and at other times there is a complete change of direction. An example of a major change was the acceptance of plate tectonics and continental drift by geologists in the 1960s, after decades of opposition. Another example is the transition from geocentricism to heliocentricism four centuries ago. And the story does not end with heliocentricity, since concepts of relativity and curved space will further alter our view of reality.

The most basic answer to the question of possible science error must be biblical. Many references declare that the truth often may be a minority position:

> *Because narrow is the gate and difficult is the way which leads to life, and there are few who find it* (Matt. 7:14).
>
> *For many are called, but few are chosen* (Matt. 22:14).
>
> *What then shall we say to these things? If God is for us, who can be against us?* (Rom. 8:31).

These verses refer to those who know their Creator personally. Although the context is not scientific truth, the verses declare how important the less popular position may be.

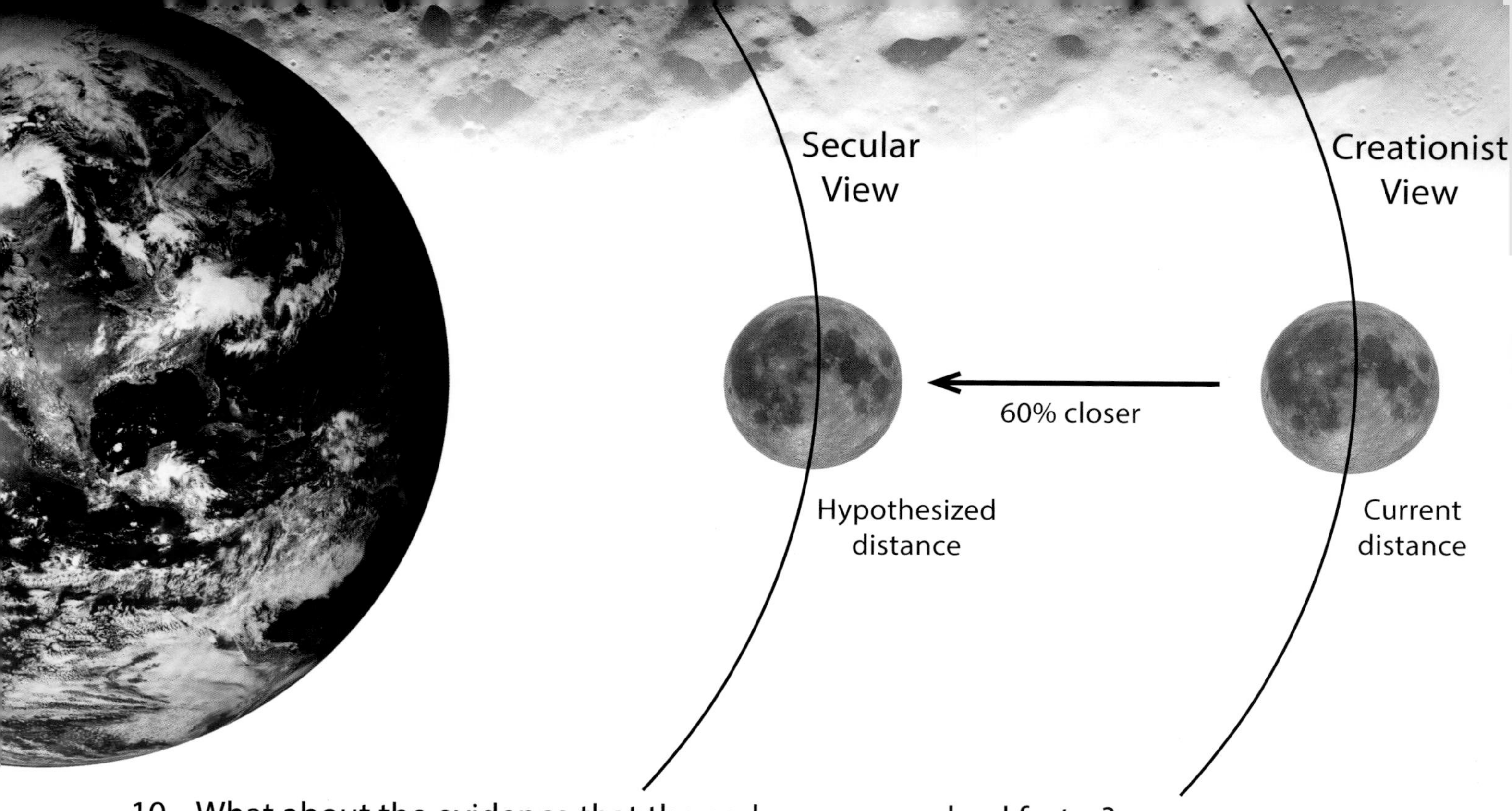

10. What about the evidence that the early moon revolved faster?

Some secular scientists have used fossil data in an attempt to "prove" that the moon was 60 percent closer to the earth 400 million years ago. Some of the fossils considered are those of the chambered nautilus. As the nautilus grows, it incorporates two repeating structures: first, new chambers in which it lives, and second, growth lines within each chamber. It is usually assumed that the growth lines occur daily, and further that a new chamber is tidally induced with each lunar cycle. If true, then the nautilus does indeed preserve a historical record of the number of days per lunar month. The number of growth lines is found to decrease sharply for fossil shells when compared with modern specimens, as few as 9 days each lunar month instead of the current 29. Thus, the conclusion is drawn that the fossils reveal short lunar months in the distant past. However, it remains an unproven hypothesis that the nautilus shell actually grows according to daily and lunar cycles. Also, the lunar records derived similarly from banding in corals and from some other *Nautiloid* species do not agree with the chambered nautilus results (Kahn, et al., 1978). Therefore, this marine fossil evidence involves a vast extrapolation with inconsistent results. In the creation view, the moon's monthly cycle remains virtually unaltered since its beginning on the fourth day of creation.

Around the Moon

One trip around the moon is equal to a round-trip flight from New York to London.

11. Are there scientific errors in Scripture?

The Bible is sometimes attacked as an untrustworthy book. If true, how then can we value its pronouncements about the moon or anything else? Upon close inspection, however, alleged Bible errors show a complete lack of insight by the critics. As one example, let us consider a supposed error involving mathematics.

In 1 Kings 7:23 a large round vessel is described, built by King Solomon around 950 B.C. as part of the temple complex in Jerusalem. Called *the Sea*, the metal container is described as 10 cubits in diameter and 30 cubits around. Now for any circle of diameter *d* and circumference *C*, the ratio C/d is the constant number, $\pi = 3.14$. However, the d and C values for Solomon's vessel give C/d = 30/10 = 3. Critics therefore claim that the Scripture dimensions give an incorrect value for pi, 3 instead of 3.14, or an error of nearly 5 percent. There are at least three possible explanations for the number difference. *First*, the numbers in 1 Kings 7:23 may be rounded off and approximate, as is commonly done with numbers. *Second*, the vessel may not have been perfectly circular. If slightly elliptical in shape, the Scripture numbers would not be expected to give the pi value exactly. *Third*, the diameter measurement may have been an outside measurement with the circumference on the inside. Suppose a cubit is 18 inches and the vessel thickness was 3 inches. Then the actual inside diameter would be 180" – 6" = 174", and the inner circumference then would be 30" x 18"= 540." The ratio then is C/d = 540"/174" = 3.1, within 1 percent of the actual value of pi.

Another example of an alleged biblical error is found in Matthew 13:32. A mustard seed is described as "smaller than all other seeds." Mustard is *not* the smallest known seed today. However, the Matthew reference further describes this seed which "a man took, and sowed in his field . . . when it has grown, it is the greatest among herbs." True to the text, mustard was indeed the smallest seed commonly used by Palestinian farmers and gardeners.

Many other challenges to Scripture accuracy likewise could be described and answered. On close inspection, all questions about accuracy have clear, simple answers. Biblical truth always remains while the criticisms fade away.

Figure 5-1. Photographs of the First Quarter (top) and Third Quarter (bottom) moons

MOON WEIGHTS
Weigh several objects and calculate their weight on the moon. You can do an online search for "weight on the moon" for various calculators.

MOON DISTANCE
At 238,866 miles from earth, how many times would one have to walk around the earth in order to reach the distance to the moon?

Moon Weight

The weight of the moon is approximately 81 quintillion tons (give or take a ton). Just one quintillion has 18 zeros so 81 quintillion would be expressed like this: 81,000,000,000,000,000,000.

CHAPTER WORD REVIEW

mediocrity principle — the assumption that there is nothing unique about the evolution of life on earth

scientific method — can be summarized in four steps: (1) understand the problem, (2) predict a solution, (3) carry out this solution, and asking (4) Is the problem solved? If not, return to step 2.

APPENDIX 1

OBSERVING THE MOON

The moon has many features that can be seen either with the unaided eye, binoculars, or a small telescope. The following list describes several areas for study. Note that most telescopes invert the moon's image, in contrast to the eye or binoculars. In general, lunar observing is best during evenings around the first and third quarter moon phases. At these times, there is a sharp contrast between lunar shadows and light. The full moon phase tends to give a washed-out appearance with less detail seen.

1. Alpine Valley — Just southeast of Crater Plato, this lava-filled valley cuts straight through the Alps Mountains. The "gash" is 1–12 miles wide and 81 miles (130 km) long. It was first noticed by Christiaan Huygens in 1727.

2. Sea of Serenity — This circular sea of basalt measures 400 miles (644 km) across. There are wrinkle ridges around the southwest edge where molten lava flowed and then hardened into basalt. The fresh crater in the south central portion is called Bessel.

3. Sea of Crises — Mare Crisium is the size of the state of Washington. Crater Bruno, discussed in chapter 2, is just beyond Crisium, around the edge on the moon's hidden side.

4. Sea of Tranquility — This region of dark basalt is where *Apollo 11* astronauts first walked in 1969. The sea area is as large as the state of Arizona.

5. Crater Tycho — This crater has long rays of bright ejected material that extend outward for hundreds of miles. At the full moon phase, the moon appears to some observers as a peeled orange with the sections meeting at Tycho. Tycho is 53 miles in diameter (85 km) and triple the depth of Grand Canyon.

6. Crater Plato — Plato is 8,000 feet (2.4 km) deep. The crater floor is very smooth and dark, and flooded with hardened lava. Many transient lunar changes have been observed here (chapter 2).

7. Sea of Rains — Mare Imbrium is so named because the moon's third quarter has long been associated with stormy weather. Just below the Sea of Rains is the Ocean of Storms (Mare Procellarum), the largest of the moon's maria.

8. Crater Archimedes — This crater, 50 miles (80 km) wide, is filled with lava, showing that it was already present when lava flooded the surrounding Mare Imbrium basin.

9. Apennines Mnts	This mountain range extends 450 miles (724 km) along the edge of the Sea of Rains (Mare Imbrium). Some of these peaks extend 20,000 feet (6096 meters-high. They may be remnants of older crater walls. During first quarter moon when the Apennines are in darkness, some of the mountaintops catch the sun and gleam as bright spots.
10. Crater Copernicus	This crater looks fresh with rays splashed across the nearby plains. The depression is about 60 miles (96 km) in diameter, formed by a colliding object several hundred feet in diameter. The thick rim has multiple rings of debris.
11. Crater Kepler	The rays surrounding this crater overlie older rock material. The crater is 20 miles (32 km) in diameter and deeper than the Grand Canyon.
12. Straight Wall	This is a hairlike line running for 70 miles (112 km) in the Sea of Clouds (Mare Nubium). It is, apparently, a crack or fault line, called by some the "railway." One side is lower than the other, forming a steep slope 1,200 feet (366 meters) high in places.

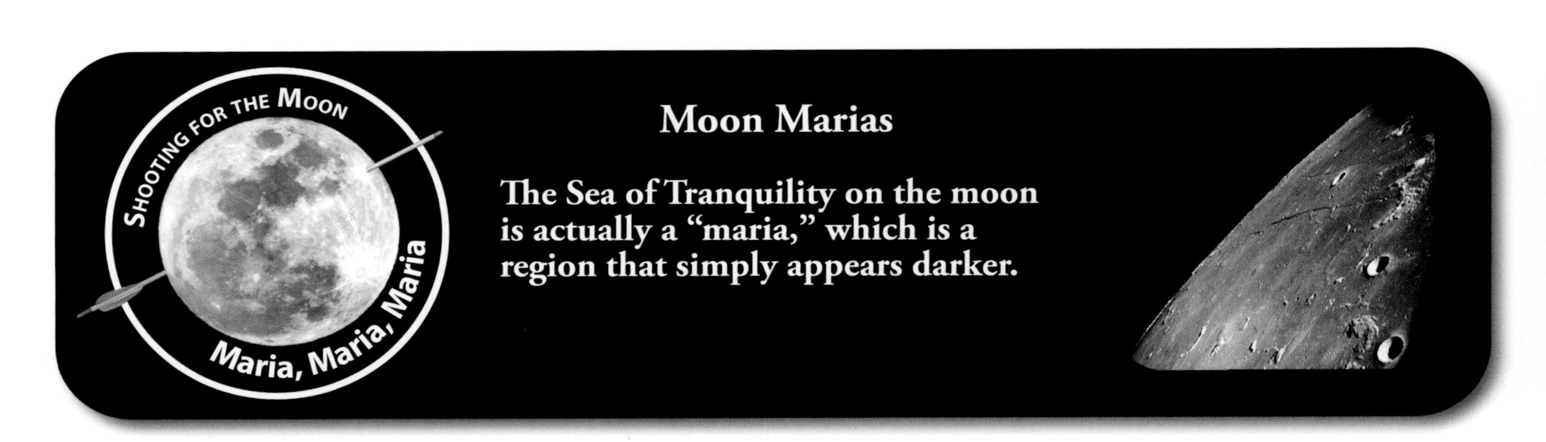

Shooting for the Moon
Maria, Maria, Maria
Moon Marias
The Sea of Tranquility on the moon is actually a "maria," which is a region that simply appears darker.

APPENDIX 2

LIVING ON THE MOON

For decades there has been an interest in colonizing our nearest neighbor, the moon. At first, the idea sounds preposterous. Who wants to live on a remote island in space, far from friends and the comforts of home? Furthermore, the moon has no air, liquid water, food, life, or protection from space radiation. Still, a lunar base provides high adventure and multiple space age benefits.

1. On earth, telescopes must look through the atmosphere, which distorts precise images. On the moon, the lack of air provides excellent telescope observing. And unlike the Hubble Telescope that drifts in space, a lunar observatory would rest on a solid surface. If placed on the hidden backside of the moon, there would be no interference from stray light or radio signals from Earth.

2. The gravity of the moon is only one-sixth that of the earth. This environment would allow long-term research into the low-gravity behavior of crystals, metals, microorganisms, medical treatments, aging, etc.

3. There is hope, perhaps overly optimistic, that frozen lakes may exist within some lunar craters. The craters in question are located near the moon's north and south poles. If water ice is indeed present, it could be melted and utilized for plant growth inside pressurized enclosures. One proposal includes large, inflatable, igloo-like structures. This could allow a community of astronauts to remain at a lunar outpost for extended periods of time.

4. Lunar water could be divided into its component elements oxygen and hydrogen, using solar energy. These are useful rocket fuels, and the moon could thus be a stepping stone for further space exploration. The oxygen could also supply essential atmosphere to living enclosures.

5. The moon is rich in metal ores. Mining these lunar resources and transporting them back to Earth might someday be economical.

6. Other benefits of lunar occupation include tourism, national security, and technological gains that inevitably follow new frontiers of research.

Beyond the physical dangers, there are political hazards to lunar colonization. Competition between countries could lead to a tense, expensive race for control of the moon. Once established there, boundary and ownership disputes could arise. Perhaps a treaty similar to that of Antarctica is possible, where no military activity is allowed and cooperative international research is encouraged.

It is unknown whether God's timetable will allow moon occupation during our lifetimes. However, He has provided mankind with the technology and ambition to pursue moon exploration. One might suggest that moon colonization is a form of creation research. God's glory is seen wherever we look, including from the vantage point of the moon. One thing is certain: if we occupy the moon, Mars, or anywhere else in space with suitcase in hand, God is already there.

GLOSSARY

accretion — the slow growth of moons or planets by the addition of small fragments, dust particles, or gas due to gravity attraction

albedo — the reflectivity of incoming light from a surface. Light and dark surfaces reveal high and low albedo, respectively. The moon reflects about 7 percent of the sunlight falling upon it.

anthropic principle — the finding that the earth, moon, and universe beyond appear to be designed for the survival and well-being of mankind

Apollo — the U.S. space program that landed six teams of astronauts on the moon during 1969–1972

asteroid — a rocky object, much smaller than a planet, that orbits the sun. Many thousands of asteroids lie between the orbits of Mars and Jupiter

astrology — an anti-scientific, cultic system that uses moon, planet, and star positions to explain and predict human actions

basalt — a common igneous rock, usually dark in color due to its iron and magnesium content. It is a fine-grained rock that cooled quickly from lava

big-bang theory — the theory that the universe began 10–15 billion years ago from the gradual expansion of an extremely dense beginning point

breccia — a rock made up of small angular fragments embedded in a fine-grained matrix

calander — Latin word meaning "to call"

capture theory — the idea that the moon was gravitationally captured by the earth

cold traps — protected spots near the moon's poles that are in permanent shadows

corona — the outer atmosphere of the sun made of hot gases

cosmogony —a theory concerning the origin of the universe and its parts

cosmology — the study of the universe including its matter, energy, origin, and destiny

crater — a circular depression resulting from a collision with a space object

creation — the supernatural origin of matter and life brought forth by the Word of God. The word *creation* also applies to the present-day universe.

crust — the solid, outermost surface layer of a moon or planet. On Earth, the crust averages about 30 miles (48 km) thick.

density — a measure of the compactness of matter. Density is determined by dividing an object's mass by its volume.

eclipse — the movement of one space object into the shadow of another. The moon is between the earth and sun during a solar eclipse. A lunar eclipse occurs at the full phase when the earth is positioned between the moon and the sun.

Enuma Elish — Mesopotamian creation myth written on seven clay tablets

evolution — the spontaneous origin of life and its development over time. An evolution world view also includes the big-bang theory and stellar evolution.

ex nihilo — a Latin term for creation meaning "out of nothing." God called material things into existence by His word.

fission theory — the idea that the moon long ago broke off from a rapidly rotating earth.

galaxy — a vast collection of about 100 billion stars held together by gravity. Galaxies are spread throughout the universe like neighboring islands. Our galaxy is the spiral Milky Way, 100,000 light years in diameter.

granite — an igneous rock composed chiefly of the colorful minerals quartz, feldspar, and biotite. Thought to have cooled slowly from molten magma.

gravity — a fundamental force of nature that causes an attraction between all objects. The moon is held captive in its orbit by the earth's gravity.

half-life — the time needed for 50 percent of remaining radioactive atoms to decay to a daughter product

harvest moon — name given to the full moon occurring nearest the autumn equinox, around September 21.

igneous — applied to rocks that form from hot, molten material (magma)

libration — the "rocking" slightly back and forth of the moon in orbit

light year — the distance through space that light travels in an entire year. This distance is about 6 trillion miles (9.5 trillion kilometers).

limb — the outer edge of the visible "disk" of the moon

lowlands — large flat areas that cover one-half of the moon's visible side

lunar — relating to the moon, from the Latin word for moon, *luna*

lunar eclipse — occurs when the earth is lined up exactly between the sun and moon

lunar highlands — rugged mountain ranges that appear from Earth as light-colored patches

lunar transient phenomena (L) — unusual lights or glowing clouds on the moon's surface, observed from the earth

magma — hot molten underground rock material that may also contain solid crystals and dissolved gasses.

mare (MAR-ay, plural *maria*) — large, dark, flat lava flows on the moon. Once thought to be seas

mass — a measure of the total amount of matter in an object. Usually measured in the metric units of grams or kilograms

mediocrity principle — the assumption that there is nothing unique about the evolution of life on Earth

meteorite — a space rock that passes through our atmosphere and strikes the earth's surface. Meteorites also strike the moon.

mineral — a naturally occurring, nonliving, crystalline solid with a unique chemical composition. One or more minerals make up a rock.

moon — any natural object that revolves around a planet as a satellite

NASA — the National Aeronautics and Space Administration, formed by the U.S. Congress in 1958

neap tides — during the first and third quarter moon phases, the sun and moon are 90 degrees apart and Earth tides then are lower.

nebula — a cloud of gas and dust in space. The plural is *nebulae.*

orbit — the path of the moon as it circles the earth, or the path of a planet around the sun. The shapes of such orbits form ellipses.

planet — an object, usually larger than 1,000 miles in diameter, that circles the sun or another star.

planetesimal — a large Mars-sized space object said by some evolutionists to have collided with Earth and formed the moon

radioactivity — the disintegration of certain unstable atoms into other kinds of atoms. High-speed particles are released in the process, called radiation.

radiosotope dating method — involves chemical analysis to determine the ratios of parent and daughter atoms in a sample.

rays — long, light-colored streaks radiating outward from craters. They consist of material "splashed" from impacts.

regolith — the surface material that overlies bedrock on the moon. Regolith is thought to be formed and stirred by countless impact collisions.

revolution — the orbital motion of one object around another. The earth revolves around the sun once each year. The moon completes a revolution around the earth in about one month.

rilles — narrow canyons or crevasses on a moon or planet's surface. They may be empty channels resulting from past lava flows.

Roche limit — breakup of a moon occurs within about 2.44 planetary radii of its host planet

rotation — the spinning or turning of an object about its own axis. The earth rotates once each 24 hours. The moon completes a rotation in about one month.

satellite — a natural or man-made object that orbits a larger object. The moon and space shuttle are satellites of the earth. In turn, the earth is a satellite of the sun.

scientific method — can be summarized in four steps (1) understand the problem, (2) predict a solution, (3) carry out this solution, and asking (4) Is the problem solved? If not, return to step 2.

sidereal period — the moon's rotation time with respect to the stars (approximately 27½ days)

solar eclipse — when the new moon moves exactly between the earth and sun

solar system — the sun and all associated objects that orbit it. These include planets, moons, asteroids, and comets.

spring tides — Tides are highest during the full moon and new moon phases when there is a lineup between the sun, earth, and moon and the gravity of both objects then contribute to the tidal pull on the earth.

synodic period — the time from one full moon to the next (approximately 29½ days)

terminator — the line separating light from darkness on the moon. At first and third quarter phases, this line divides the moon in half.

theistic evolution — the idea that God created all things through evolutionary process. The big-bang theory, long ages, and the animal-to-human transition are all accepted.

vernal equinox — the first day of spring, around March 20–22.

yom — Hebrew word translated "day"

REFERENCES

Balling, R.C., and R.S. Cerveny. 1995. Influence of lunar phase on daily global temperatures. *Science* 267:1481–1482.

Beek, Martin. 1962. *Atlas of Mesopotamia.* New York: Nelson.

Calame, O., and J.D. Mulholland. 1978. Lunar crater Giordano Bruno: a.d. 1178 impact observations consistent with laser ranging results. *Science* 199:875–877.

Cherrington, Ernest H. 1969. *Exploring the Moon through Binoculars.* New York: McGraw-Hill.

Comins, Neil F. 1991. The earth without the moon. *Astronomy* 19(2):40–55.

Comins, Neil F. 1993. *What If the Moon Didn't Exist?* New York: Harper Collins Publishers, Inc.

Corliss, William R. 1975. *Strange Universe: A Sourcebook of Curious Astronomical Observations.* Glen Arm, MD:Sourcebook Project.

Davis, John D., and John C. Whitcomb. 1989. *Israel: From Conquest to Exile.* Winona Lake, IN: BMH Books.

DeYoung, Don B. 1990. The earth-moon system. *Proceedings of the Second International Conference on Creation.* Vol. II. p. 79–84.

DeYoung, Don B. 2000. Lunar crater Giordano Bruno. *Creation Research Society Quarterly* 37:185–188.

Hooykaas, Reijer. 1972. *Religion and the Rise of Modern Science.* Grand Rapids, MI: Wm. B. Eerdmans Publishing Co.

Humphreys, D. Russell. 1994. *Starlight and Time.* Green Forest, AR: Master Books.

Kahn, P.G.K., and S.M. Pompea. 1978. Nautiloid growth and dynamical evolution of the earth-moon system. *Nature* 275:606–611.

Kuhn, Thomas. 1970. *The Structure of Scientific Revolutions.* Chicago, IL: University of Chicago Press.

Ley, Willy. 1965. *Ranger to the Moon.* New York: McGraw-Hill.

Lissauer, Jack. 1997. It's not easy to make the moon. *Nature* 389:327–328.

Mitroff, I.I. 1974. *The Subjective Side of Science: A Philosophical Inquiry into the Psychology of the Apollo Moon Scientists.* Amsterdam: Elsevier Science Publishing Co.

Moorhead, Paul S., and Martin M. Kaplan (eds.). 1967. *Mathematical Challenges to the Neo-Darwinian Interpretation of Evolution.* Philadelphia, PA: Wistar Institute Press.

Pearce, Fred. 1997. A full moon warms icy wastes. *New Scientist* 153:15.

Pritchard, James B., editor. 1969. *Ancient Near Eastern Texts Relating to the Old Testament*. Princeton, NJ: Princeton University Press. The full text of *Enuma Elish* is on p. 60–68.

Ross, Hugh. 1994. *Creation and time*. Colorado Springs, CO: NavPress Publishing Group.

Snoke, David. 2001. In favor of God-of-the-gaps reasoning. *Perspectives* 53(3):152–158.

Sheeham, William, and Thomas Dobbins. 2001. *Epic Moon*. Richmond, VA: Willmann-Bell, Inc.

Thiele, Edwin R. 1965. *The Mysterious Numbers of the Hebrew Kings*. Grand Rapids, MI: Eerdmans.

Ward, Peter D., and Donald Brownlee. 2000. *Rare Earth.* New York: Springer-Verlag.

Weiss, Peter. 2001. Constant changes. *Science News* 160(14):222–223.

Whitcomb, John C. 1977. *Chart of Old Testament Kings and Prophets.* Winona Lake, IN: BMH Books.

Whitcomb, John C., and Don B. DeYoung. 1978. *The Moon: Its Creation, Form, and Significance.* Winona Lake, IN: BMH Books.

Wunsch, Carl. 2000. Moon, tides, and climate. *Nature* 405:743–744.

INDEX

ABOUT THE AUTHORS

Dr. Don DeYoung is on the math/science faculty of Grace College, Winona Lake, Indiana. His writing goal is to popularize and clarify science topics including connections with the creation worldview. Several of his 16 books feature a Question-Answer approach with concise discussion of popular issues. Don enjoys backpacking, running, speaking on the Bible and science, and travel with his wife, Sally.

Before his passing in early 2020, **Dr. John C. Whitcomb** was president of Whitcomb Ministries, Inc., and founder and professor of Christian Workman Schools of Theology. He and his wife Norma resided in Indianapolis and enjoyed spending time with their 6 children and 17 grandchildren.

Dr. Whitcomb had been a professor of Old Testament and theology for more than 50 years and was widely recognized as a leading Biblical scholar. Norma Whitcomb is a Bible teacher and seminar speaker to ladies in the United States and abroad.

Dr. Whitcomb taught at Grace Theological Seminary in Winona Lake, Indiana, from 1951 to 1990, and gained much recognition for his work on *The Genesis Flood*, which he co-authored with Dr. Henry Morris in 1961. This book has been credited as one of the major catalysts for the modern Biblical creationism movement.

Dr. Whitcomb's life and ministry may be summed up in this quotation: "I want to be in the full-time business of finding out what God says and telling as many people as I can."

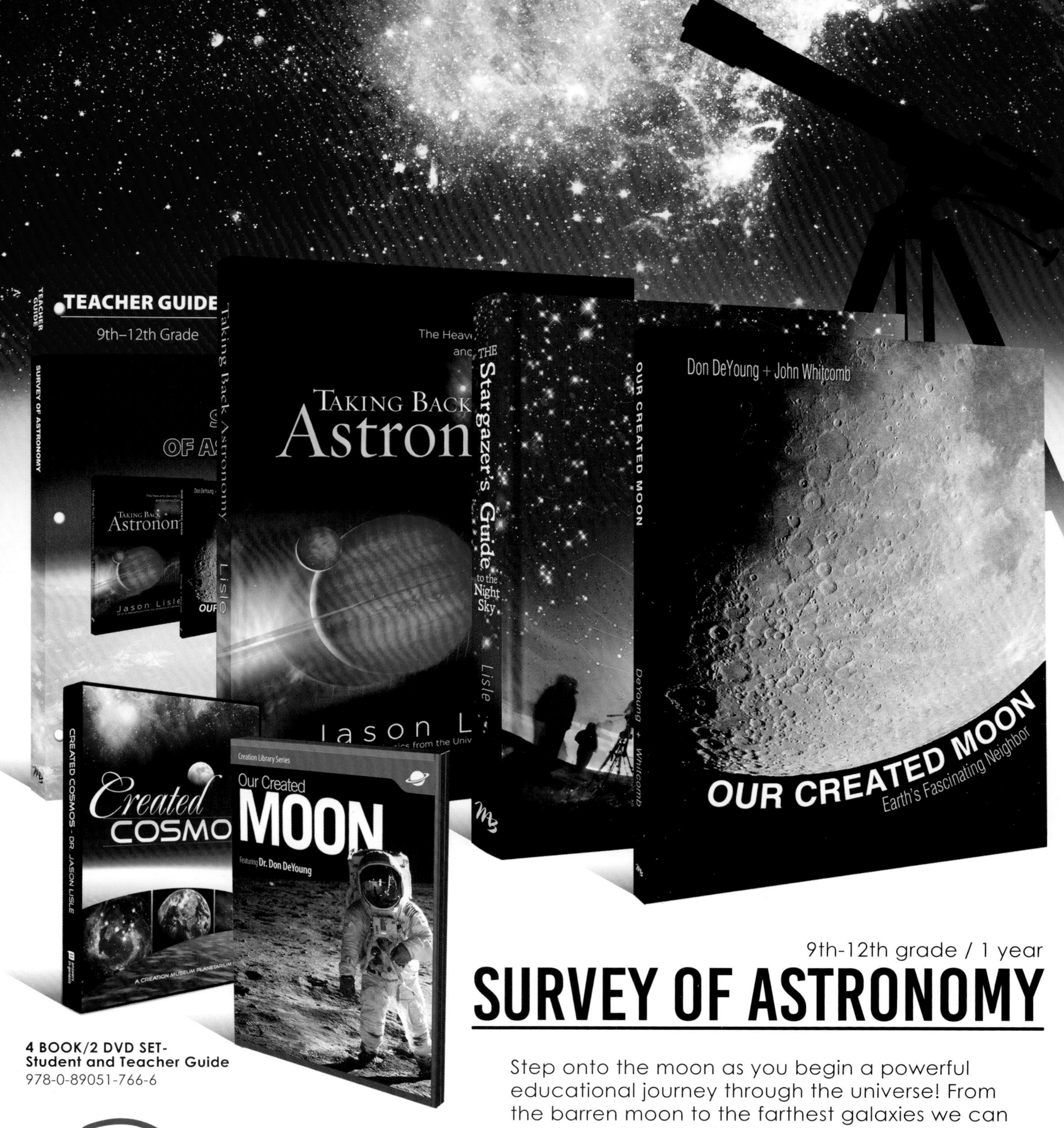
TEACHER GUIDE
9th–12th Grade
SURVEY OF ASTRONOMY
Taking Back Astronomy
THE Stargazer's Guide to the Night Sky
Lisle
OUR CREATED MOON
Don DeYoung + John Whitcomb
OUR CREATED MOON
Earth's Fascinating Neighbor
DeYoung + Whitcomb
CREATED COSMOS - DR. JASON LISLE
Creation Library Series
Our Created
MOON
Featuring Dr. Don DeYoung

MASTERBOOKS
CURRICULUM

Daily Lesson Plan

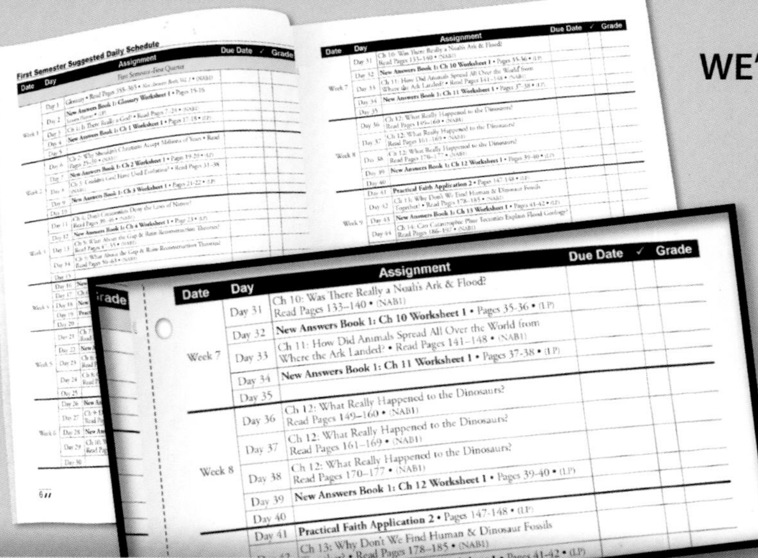

Date	Day	Assignment	Due Date	✓	Grade
Week 7	Day 31	Ch 10: Was There Really a Noah's Ark & Flood? Read Pages 133–140 • (NAB1)			
	Day 32	**New Answers Book 1: Ch 10 Worksheet 1** • Pages 35-36 • (LP)			
	Day 33	Ch 11: How Did Animals Spread All Over the World from Where the Ark Landed? • Read Pages 141–148 • (NAB1)			
	Day 34	**New Answers Book 1: Ch 11 Worksheet 1** • Pages 37-38 • (LP)			
	Day 35				
Week 8	Day 36	Ch 12: What Really Happened to the Dinosaurs? Read Pages 149–160 • (NAB1)			
	Day 37	Ch 12: What Really Happened to the Dinosaurs? Read Pages 161–169 • (NAB1)			
	Day 38	Ch 12: What Really Happened to the Dinosaurs? Read Pages 170–177 • (NAB1)			
	Day 39	**New Answers Book 1: Ch 12 Worksheet 1** • Pages 39-40 • (LP)			
	Day 40				
	Day 41	**Practical Faith Application 2** • Pages 147-148 • (LP)			

WE'VE DONE THE WORK FOR YOU!

PERFORATED & 3-HOLE PUNCHED

FLEXIBLE 180-DAY SCHEDULE

DAILY LIST OF ACTIVITIES

RECORD KEEPING

"THE TEACHER GUIDE MAKES THINGS SO MUCH EASIER AND TAKES THE GUESS WORK OUT OF IT FOR ME."

Master Books® Homeschool Curriculum

Faith-Building Books & Resources
Parent-Friendly Lesson Plans
Biblically-Based Worldview
Affordably Priced

Master Books® is the leading publisher of books and resources based upon a Biblical worldview that points to God as our Creator.

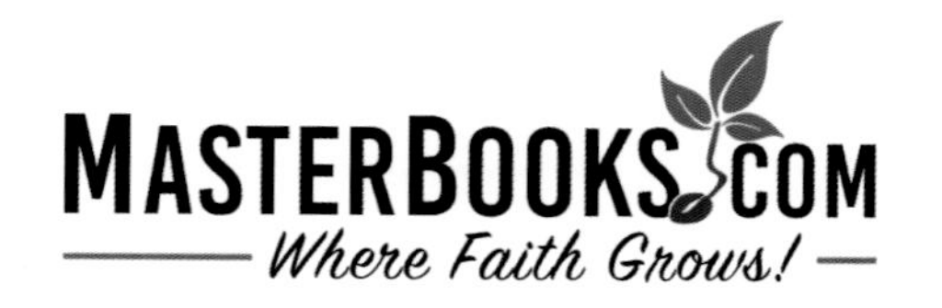